A Legacy of Change

A Legacy of Change

HISTORIC HUMAN IMPACT ON VEGETATION
IN THE ARIZONA BORDERLANDS

Conrad Joseph Bahre

THE UNIVERSITY OF ARIZONA PRESS   TUCSON

The University of Arizona Press
Copyright © 1991
The Arizona Board of Regents
All Rights Reserved

This book was set in 10/13 Linotype CRT Trump.
⊚ This book is printed on acid-free, archival-quality paper.
Manufactured in the United States of America.

96 95 94 93 92 91   6 5 4 3 2 1

Library of Congress Cataloging-in-Publication Data
Bahre, Conrad J.
   A legacy of change : historic human impact on vegetation in the Arizona borderlands / Conrad Joseph Bahre.
     p.   cm.
   Includes bibliographical references and index.
   ISBN 0-8165-1204-3
   1. Man—Influence on nature—Arizona. 2. Land use—Environmental aspects—Arizona. 3. Landscape assessment—Arizona. 4. Climatic changes—Arizona. 5. Vegetation dynamics—Arizona. I. Title.
GF504.A6B34   1991
581.5′264′0979153—dc20   90-39777
                                                                                              CIP

British Library Cataloguing in Publication data are available.

To my mother
Rosemary Johantges Bahre (1918–1989)

Contents

List of Illustrations	ix
List of Tables	xi
Foreword	xiii
Acknowledgments	xvii

Part I. Historic Vegetation Change

1. Introduction	3
2. Setting	18
3. Previous Studies of Vegetation Change	42
4. Assessing Vegetation Change	59

Part II. Primary Historic Human Impacts Since 1870

5. Livestock Grazing	109
6. Fire	124
7. Fuelwood Cutting	143
8. Exotic Plant Introductions	155
9. Agriculture, Logging, and Haying	161
10. Patterns and Factors of Change	176
11. Summary and Conclusions	185
Appendix A. Selected Section Lines with Oaks Recorded by the Public Land Survey in Southeastern Arizona	188

Appendix B. Tree and Large Woody Shrub Counts
　　　　　　　Along 111 Matching Transects on
　　　　　　　SCS (1935–1937) and NHAP (1983–1984)
　　　　　　　Imagery by Vegetation Type　　　　　　　　189

Appendix C. Z-Scores for the Total Sample　　　　　　192

Notes　　　　　　　　　　　　　　　　　　　　　　　195

Bibliography　　　　　　　　　　　　　　　　　　　　197

Index　　　　　　　　　　　　　　　　　　　　　　　221

Illustrations

1.1	Place-Name Maps of Southeastern Arizona	4
2.1	Average Annual Precipitation in Southeastern Arizona	19
2.2	Vegetation Cover of Southeastern Arizona	21
2.3	Sonoran Desertscrub	22
2.4	Chihuahuan Desertscrub	23
2.5	Semidesert Grassland	24
2.6	Plains Grassland	25
2.7	Evergreen Woodland	26
2.8	Rocky Mountain (Petran) Conifer Forest	27
2.9	Riparian Wetland	28
2.10	Confirmed Spanish and Mexican Land Grants in Southeastern Arizona	33
3.1	Landsat Image of the Arizona-Sonora Borderlands	50
4.1	Sonoran Desertscrub Along the Western Section Line of S20, T13S, R16E, near Bellota Ranch, East of Tucson	61
4.2	Contention	68
4.3	Fairbank	70
4.4	"Hill of San Cayetano"	72
4.5	Greaterville	74
4.6	Sonoita Creek	76
4.7	Arizona-Pittsburg Mine	78
4.8	Aravaipa Canyon	80
4.9	Rosemont Mine	82

x   Illustrations

4.10	Pusch Ridge	84
4.11	Santa Rita Experimental Range	86
4.12	Monument 111	88
4.13	Matching SCS (1936) and NHAP (1984) Aerial Photographs of Oak-Grassland near Nogales	92
4.14	Matching SCS (1936) and NHAP (1983) Aerial Photographs of Oak-Grassland near Sonoita	94
4.15	Histogram of Absolute Change in Tree and Large Woody Shrub Numbers by Transect Between 1935 and 1984	96
4.16	Histogram of Percent of Change in Tree and Large Woody Shrub Numbers by Transect Between 1935 and 1984	97
4.17	Ten-year Running Means Showing Mean Seasonal Precipitation at Eighteen Southern Arizona and Western New Mexico Stations	101
4.18	Annual Deviation from Mean Precipitation for Nineteen Arizona Stations: 1898–1986	103
4.19	Annual Deviation from Mean Precipitation for Four Stations in the Study Area: 1898–1986	104
5.1	Cattle Roundup in the Sulphur Springs Valley circa 1890	110
6.1	Lightning Versus Anthropogenic Wildfires in the Coronado National Forest, 1959–1986	126
6.2	Lightning Fire Occurrence by Elevation in the Rincon Mountains, 1937–1986	126
6.3	H. M. Robert's Map of Southeastern Arizona in 1869	136
7.1	Ricks of Cordwood at the Contention Mill, Contention, 1882	146
8.1	Major Areas of Lehmann Lovegrass in Southeastern Arizona	159
9.1	Irrigated Cropland Along the Santa Cruz River in Tucson, 1882	163
9.2	Ross Sawmill in Morse Canyon in the Chiricahua Mountains circa 1885	167
9.3	Stacks of Wild Hay on the Santa Rita Experimental Range, 1902	173

Tables

2.1 Settlement and Population of Southeastern Arizona in 1860, 1870, and 1880 — 35

3.1 Relationships Between Disturbance and Stream Entrenching in Southeastern Arizona — 45

3.2 Proposed Vegetation Changes in Southeastern Arizona Since 1870 — 54

3.3 Proposed Primary Cause of Vegetation Changes in Southeastern Arizona Since 1870 — 57

4.1 Sites, Geographic Coordinates, Negative Numbers in the Hastings-Turner Collection, Vegetation Types, and Elevations of the Ground Photographic Stations — 64

4.2 Z-Scores for the Total Sample and for the Total Sample Minus Chihuahuan and Sonoran Desertscrub — 98

4.3 Variability of Annual Precipitation — 102

4.4 Drought and Wet Periods: Cumulative Precipitation Deficiency or Excess Between June and September at Fort Lowell, near Tucson — 102

6.1 Coronado National Forest Fire Statistics, 1959–1986 — 127

6.2 Wildfire Frequency in the Major Vegetation Types of Southeastern Arizona, 1859–1890 — 130

7.1 Domestic Fuelwood Consumption for the Tombstone Mining District During the Tombstone Silver Bonanza (June 1879–Dec. 1886) — 149

7.2 Estimated Silver Ore Production and Stamp Mill
Fuelwood Consumption During the Tombstone
Silver Bonanza (June 1879–Dec. 1886)   150

8.1 Exotic Plant Varieties Recommended for Release
by the Tucson Plant Materials Center of the Soil
Conservation Service in 1987   157

Foreword

Attempting to unravel the relationships between humans and the landscape, and the ways in which they interact, is a basic and traditional task of the geographer. For reasons that are unclear, however, this line of research recently has been largely—and often vigorously—ignored by geographers, primarily in the United States. Aside from a growing but still small number of anthropologists, members of other disciplines have not pursued human-land research themes.

A Legacy of Change is witness to the value of this tradition. By examining historical land uses and the impact they have had on "natural" vegetation in southeastern Arizona, Conrad Bahre shows how poorly understood is the relationship between human activities and vegetation in a well-studied part of the United States. Much of what is presented here is new, but much is also a reinterpretation of data drawn from other works. As suggested in the title, Bahre rejects theories of vegetation change based exclusively on climate. Given the evidence, his conclusions are almost embarrassingly obvious, yet they have been met with a surprising skepticism by students of vegetation change within the region. Undoubtedly, his findings will fuel continued debate over the reality of recent climate change, especially in Arizona. But, beyond its contribution to local natural history, the major points raised in *A Legacy of Change* transcend regional debate as we struggle to understand change on a global scale.

Since 1985, the earth sciences have undergone a revolution of sorts. Today, almost all major research efforts concern themselves to varying degrees with problems extracted from the broad global change agenda. It is not surprising, therefore, that much attention in this expansive arena has come to rest on the changes produced directly or indirectly by human actions. More specific to our discussion, a good part of this interest has gravitated to land surface–atmosphere interactions, especially in regard to changes in vegetation cover and their relationship with climate.

Vegetation has always held a prominent and peculiar place in the study of climate for two related reasons. First, confronted with the problem of describing climate over large areas based on a very few observation points, climatologists have sought other more easily observed indicators that integrate local precipitation and temperature. Vegetation is commonly viewed as the most obvious integrating product of climate and is undoubtedly the most familiar and widely accepted climate surrogate. Thus, rather than relying on sparse station data, almost all maps of global or regional climate are derived to some degree from unambiguous vegetation patterns.

Second, in attempting to detect and document climate variation, we are constrained by instrumental climatological records that seldom extend more than a century into the past. Obviously, this is an extremely slender base upon which to build and validate models that might tell us about the future. Not surprisingly, the development of a long-term climatic database has emerged as one of the most pressing and challenging tasks in the global change effort.

In fact, the absolute necessity of extending the climatic record back in time by using proxy climatic data has spawned a number of subdisciplines such as dendrochronology, palynology, and packrat midden analysis, all of which are used worldwide in reconstructing past climates. As might be suspected, many of the tools employed in these efforts are based on the close relationship between vegetation and climate. Stated simply, it is assumed that over time, changes in climate will be reflected in changes in vegetation; conversely, changes in vegetation are the result of changes in climate.

Bahre directly challenges this approach when applied to the recent past by pointing out that, at least for southeastern Arizona, the knowledge of vegetation dynamics appears to be fundamentally flawed.

Previous interpretations of vegetation change in this region fail for at least three reasons. First, there has been a consistent and persistent underestimation of the human factor in vegetation dynamics. As Bahre documents, the ways in which man manipulates and shapes vegetation are diverse and often subtle, but are ultimately knowable. The lack of sensitivity to human agency is in many ways peculiar, particularly with the wave of attention given the role of man in the recent "desertification" of large areas of the African Sahel, or the more distant and conjectural destruction of the Rajasthani environment in India and Pakistan. Moreover, there are similarities between territorial southern

Arizona in the 1890s and the Sahel of the 1970s (overgrazing terminated by severe drought) that are hardly overdrawn and that have not been explored comparatively.

A second failure is in the identification and interpretation of change. Initial conditions (those at the beginning of the period of record) may not represent equilibria. Moreover, they may not even be representative of conditions in general. As Bahre emphasizes, much of the evidence of change in southeastern Arizona is drawn from repeat photography, seldom spanning a period of more than 100 years. As shown here, presumed unequivocal evidence of change found in the photographic record may not, in fact, be real. Errors in interpretation of change appear to result from an incomplete understanding of historic land-use activities, an assumption that original samples are unbiased and representative of general conditions, or both. Either of these errors may lead to incorrect interpretation of initial conditions.

Third, and perhaps most importantly, is that vegetation dynamics are also controlled by the coincidence of random events, all of which may play critical roles. This conclusion is not new, but, until now, the diversity and magnitude of the human contribution to change was not adequately assessed.

The "invasion" of woody plants in the grasslands of southern Arizona is perhaps the best example of convergent independent events explained by Bahre. As revealed in long-term proxy data, the sequence of extreme climatic events around the turn of the century (severe drought followed by significant rainfall) is not unprecedented. Yet, until the late 1890s, these sequences had little lasting effect on the landscape. At that time, however, severe drought, preceded and accompanied by overgrazing, followed by increased storm events coupled with fire suppression, made for conditions under which divergent factors could create a favorable environment for the establishment and maintenance of woody plants in what was previously grassland. Although not the only cause of change, human intervention seems to have been the critical precipitating factor.

A Legacy of Change has already caused some controversy in the hothouse of vegetation change research in southern Arizona by questioning whether or not change has actually occurred in the region. But, as already suggested, this ignores larger questions—both methodological and hypothetical—regarding the role of humankind as an agent of regional change.

From a classically geographic perspective, the arguments put forward by Bahre suggest that, instead of doggedly seeking natural causes of change, a prudent working hypothesis must first assume that change is a product of human action. Given the weight of evidence presented here, it would seem that only after human agency is understood and accommodated can we begin to consider other explanations for landscape change.

If Bahre is correct in his arguments, the ramifications are profound for reconstructions of recent climate based on vegetation. There are few areas in the arid and semiarid zone that are better known than southeastern Arizona. If the evidence for short-term change is so elusive and interpretation so contentious here, how confident can we be of our ability to understand recent climate history in other lesser known regions of the world?

CHARLES F. HUTCHINSON

Acknowledgments

Many people have contributed to this book. Ray Turner made the repeat photography in the Hastings and Turner Collection available to me, and reviewed the manuscript. Bob Humphrey, Tom Whitlow, Mitch McClaren, and Tom Swetnam read all or parts of the book and made many valuable suggestions. David Lyster collected the Soil Conservation Service aerial photos at the National Cartographic Archives in Washington, D.C., and Tina Kennedy helped me obtain the National High Altitude Photography from the Agricultural Stabilization and Conservation Service in Denver. Steve Jennings gathered much of the information from nineteenth-century newspapers at the Bancroft Library, University of California at Berkeley, helped review the U.S. Public Land Survey surveyors' field notes at the Bureau of Land Management in Phoenix, designed the sampling methods for the transects, and did the statistical work. William Sellers furnished me with his data sets of Arizona's climate, and Marlyn Shelton aided in the analysis of those data. Stuart Allan of Allan Cartography in Medford, Oregon, did the place-name map of southeastern Arizona, and Mary Cunha did most of the other cartography. John Turner supplied information on fires, fuelwood cutting, and grazing in the Coronado National Forest. Heidi Seney did the editing; Maria Hope did the typing; and Barbara Beatty encouraged me to send the manuscript to the University of Arizona Press. To all of these people I am particularly indebted.

Others who have contributed in one way or another to the development and completion of the book are Michael Barbour, Jonathan Sauer, Tom Vaughn, Brita Mack, S. Clark Martin, Bill Piper, Lee Poague, Dave Bradbury, Jerry Cox, Tony Burgess, Jan Bowers, Glenn Goodman, Ruth Tilton, Joe Escapule, Chuck DiPeso, Paul S. Martin, Dean Lippert, Bill Brown, Marvin Garcia, Chuck Ames, Chris Baisan, Dutch Martinich, Stone Collie, Ariel Appleton, Jane Bock, Jill Earick, Buster Pyeatt, Gary Nabhan, Gary Newman, Howard Boice, Bob Oglesby, Mike Winter,

Mike Parton, Dick Reeves, Ted Knipe, Fran Ramos, Ken Bennett, Tom Harlan, Alex McCord, Bruce Munday, Marshall Ashburn, Richard Harrison, John Tucker, Peter Warren, George and Bessie Bercich, Stan Clemmons, Tom Sheridan, Julio Betancourt, George and Sissy Bradt, Pat Spoerl, Fred Gehlbach, and Rosamel de la Osa.

I also wish to thank the staffs of the Special Collections Library at the University of Arizona, Bancroft Library, Huntington Library (San Marino, Calif.), Arizona Historical Society, Bisbee Mining and Historical Museum, Bureau of Land Management, and the Coronado National Forest, all of whom expedited my research.

Finally, I wish to extend special thanks to Duncan and Kathleen Taylor and to Chuck and Barbara Hutchinson, who made their homes available to me during long stays in Tucson. In addition, both Duncan and Chuck helped with the field work. Chuck and I had originally decided to co-author this book in 1981, but his administrative commitments got in the way. Nevertheless, Chuck had a great deal to do with the development of the methodology, accompanied me on many field trips, and helped in the preparation of the final manuscript. He also wrote the foreword. To Chuck I am particularly grateful.

Financial support for this study was provided by a series of Faculty Research Grants from the University of California at Davis. Without that support the completion of this book would have been next to impossible. I particularly wish to thank Professor Robert Powell, who managed to obtain monies for me to complete the typing of the manuscript.

PART I
Historic Vegetation Change

1 Introduction

In 1955 the Wenner-Gren Foundation sponsored an international symposium, "Man's Role in Changing the Face of the Earth," dedicated to the American statesman and scholar George Perkins Marsh (Thomas 1956). Marsh, who wrote two books detailing the ways in which man has modified the earth—*Man and Nature* (1864) and *The Earth as Modified by Human Action* (1874)—denounced environmental determinism, a prominent theme in geography and history in the nineteenth century, and pointed out that human impacts on the environment are neither negligible nor necessarily benevolent. By the end of the nineteenth century the subject of man as a dominant agent in modifying the natural environment, especially its vegetation cover, became an important theme of geographic investigation.[1]

In the United States, geographic research concerning human impacts on vegetation and vegetation change is best represented by the "Berkeley school of geography" founded in 1923 by Carl Sauer, who championed the cultural-historical approach to landscape evolution. In his view the modern vegetation cover of the inhabited world is the result not only of natural evolutionary processes but also of historic human perceptions and uses of the land.[2]

In spite of the contributions of Sauer and his students, little research has been done on how historic land uses have affected the evolution of North America's wild landscapes. This is true of the American Southwest, particularly of southeastern Arizona (see Figure 1.1), where directional vegetation changes since the mid-nineteenth century—such as the taking over of rangelands by woody plants—are often attributed to a naturally recurring stress—drought—or to a recent trend toward greater aridity.

Except for grazing and wildfire suppression, the effects of most historic land uses on vegetation have not been adequately identified, nor have the land uses themselves been analyzed in detail. This is remarkable because most documented vegetation changes occurred after major Anglo-American settlement of the region in the early 1870s.

Before 1870, according to historical record, the biological environment of the semiarid basin and range country of southeastern Arizona was comparatively lush. In the 1850s and 1860s, grass was plentiful; the grasslands were open and fairly brush-free; the rivers had perennial flow throughout much of their courses and were in parts unchanneled and lined with galeria forests of willows and cottonwoods; marshes (ciénagas) and stands of mesquite and sacaton covered large areas of bottomland; wildfires were fairly common; the ponderosa pine and mixed-conifer forests had uneven-aged stands and open understories; malaria was rampant in certain areas; fish were plentiful; and antelope, prairie dog, grizzly bear, otter, beaver, and wolf were abundant.

Today, the picture is different. Native grasses have declined and have been replaced in part by exotics; many areas of grassland have been taken over by mesquite, acacia, burroweed, and snakeweed; the rivers have entrenched themselves and in parts have dried up; the native riparian vegetation has largely disappeared or has been replaced by exotics; wildfires have declined in frequency; the ponderosa pine and mixed-conifer forests are even-aged and have more intense fires than before; much native vegetation has been cleared for settlement; malaria has disappeared; and all of the animals mentioned, except for a few introduced antelope, have been wiped out. These changes did not occur everywhere at the same time, but they are related and share at least one common denominator—the advent of Anglo settlement.

Most of the evidence for vegetation change in southeastern Arizona since 1870 has been documented in different subsets of historical-modern matched or repeat ground photographs (Hastings and Turner 1965;

FIGURE 1.1. Place-name maps of southeastern Arizona. The index map (page 5) is followed by eight large-scale sectional maps of the study area. Maps 1–4 cover the northern part of the study area from west to east. Maps 5–8 cover the southern part from west to east.

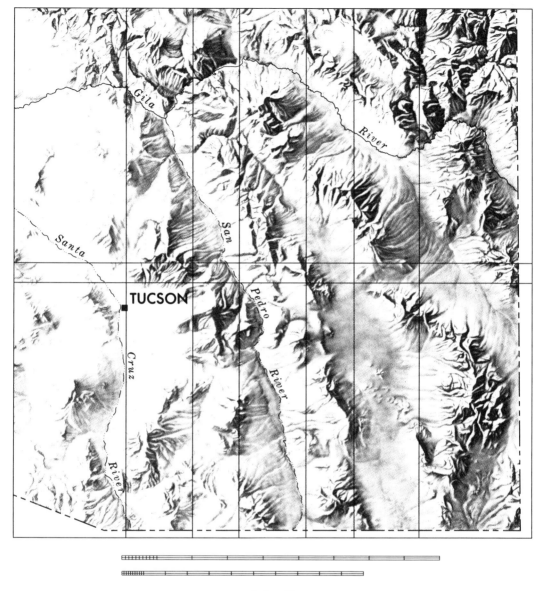

Index Map

Scale for Maps 1–8 following:

1:550,000

1 inch = approximately 9 miles
1 centimeter = 5.5 kilometers
Elevations in feet

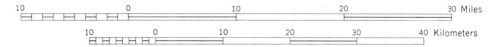

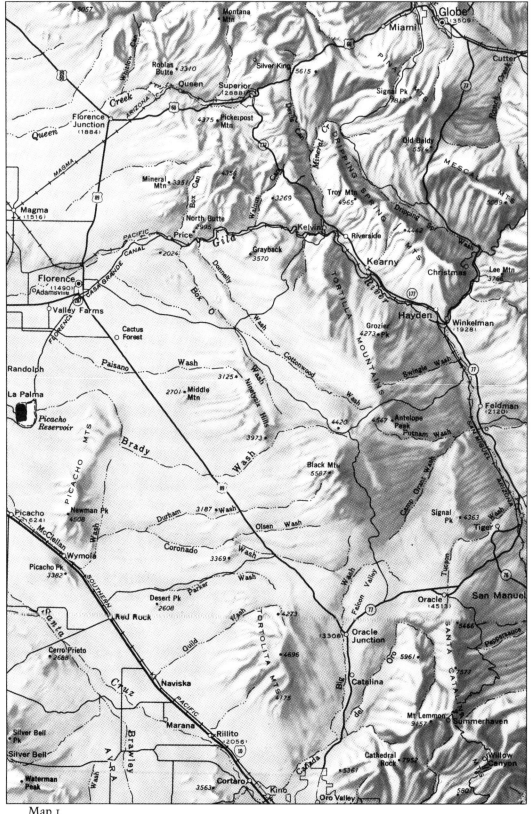

Map 1

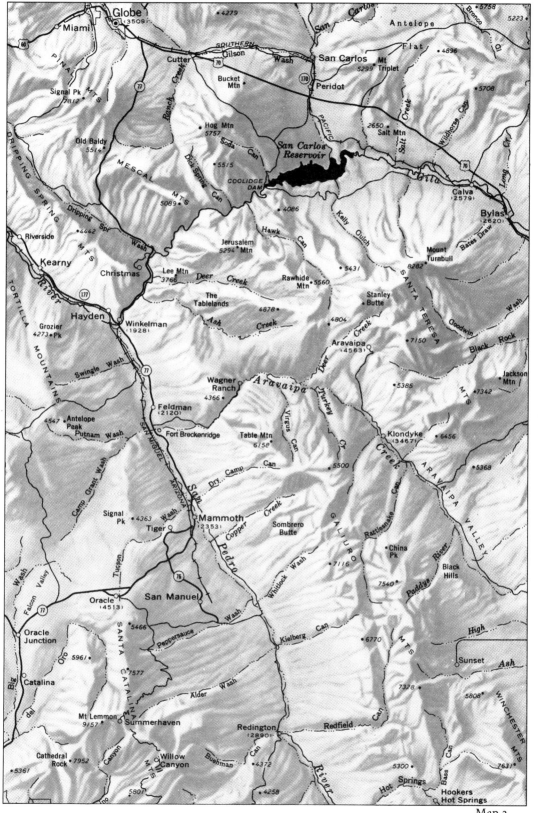

Map 2

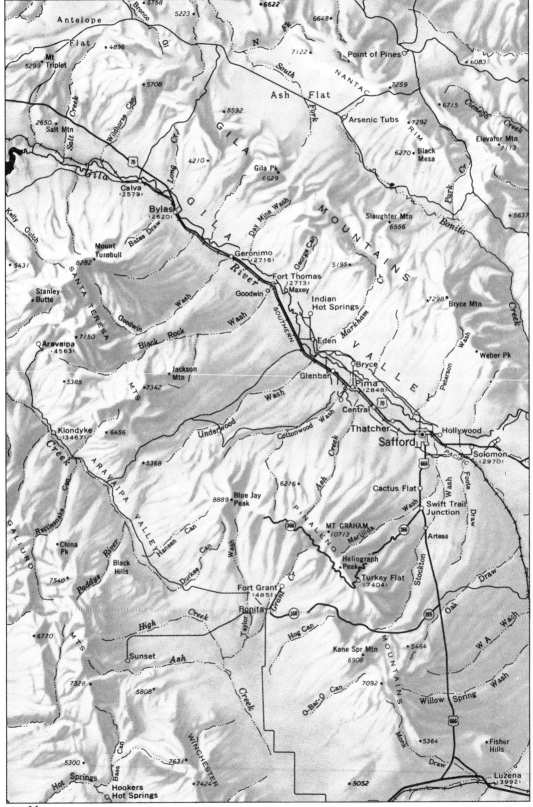

Map 3

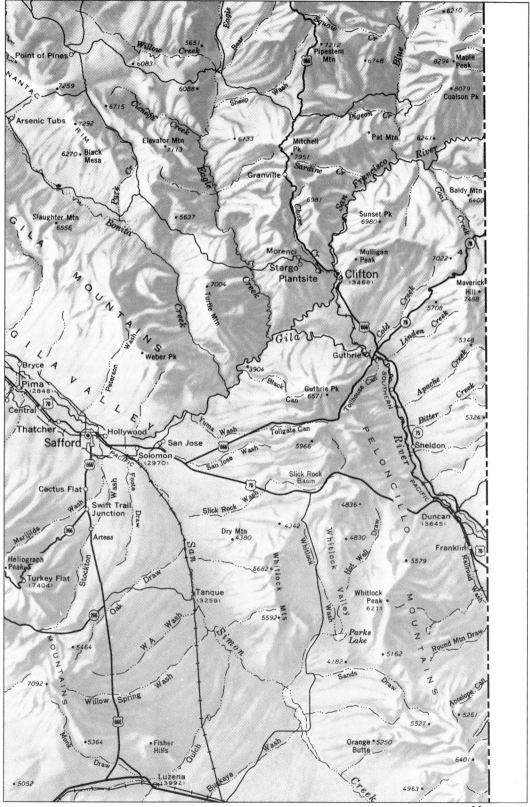

Map 4

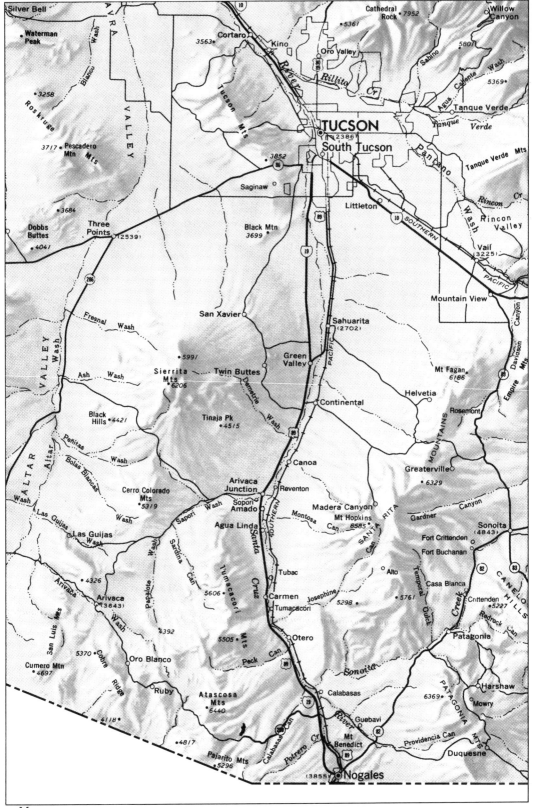

Map 5

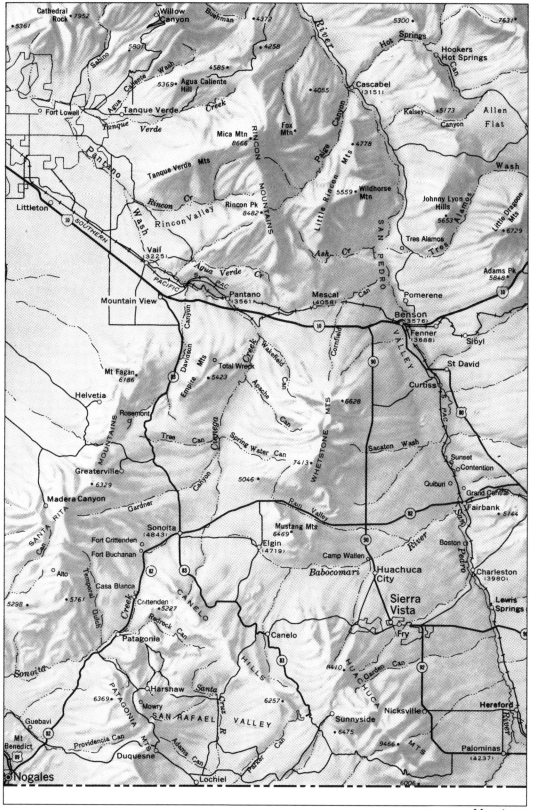

Map 6

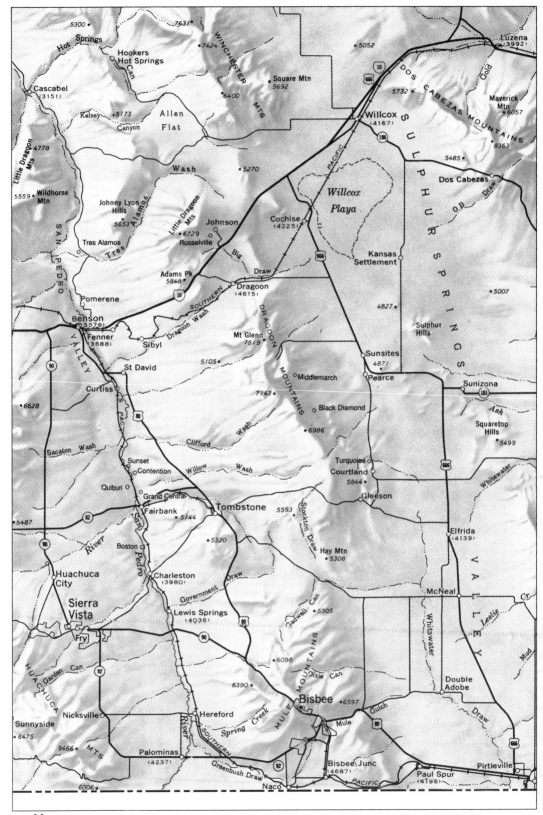

Map 7

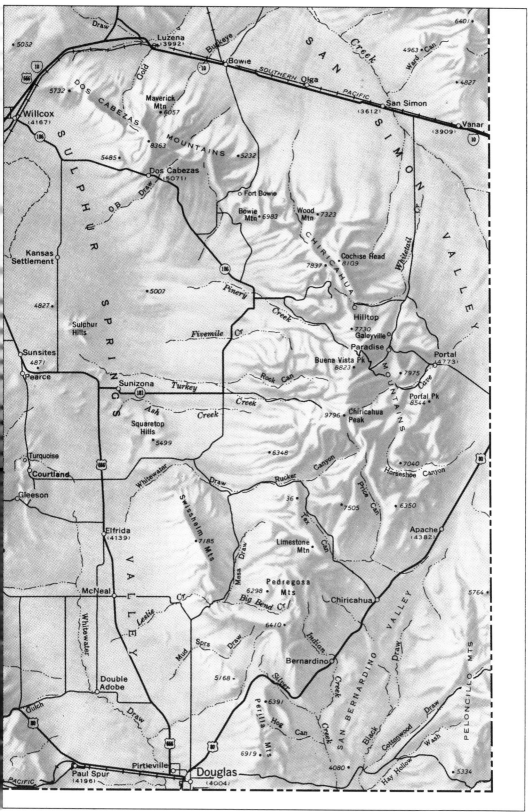

Map 8

Martin and Turner 1977; Bahre and Bradbury 1978; Gehlbach 1981; Humphrey 1987). There are, however, several drawbacks in using the changes in these photographs as evidence for directional vegetation changes and regional shifts in plant ranges or in claiming that the changes are largely the result of natural factors.

1. The present collections of matched-ground photographs cover little of the total area.
2. The patterns of change are random and inconsistent. For example, one pair of matched photographs shows a decline in sahuaro numbers at one site since the 1920s, while other pairs taken from a nearby camera station, as well as at distant sites at similar elevations, show no change or an increase in sahuaro numbers during the same time span.
3. The field of view in ground photographs is usually oblique and narrow, which limits their usefulness for determining changes in plant distribution.
4. Most of the matched photos are of disturbed sites.
5. Because no land-use histories of any of the photographed sites exist and little is known about the effects of land uses on the evolution of modern vegetation, it usually cannot be determined which changes are caused by natural and which by anthropogenic factors, and whether the changes are directional, cyclic, or simply representative of revegetation of disturbed sites.
6. Most of the earliest photographs were taken a decade or more after the sites were first disturbed by major Anglo land uses.

The hypotheses presented to explain vegetation changes since 1870 fall into three categories: (1) those which maintain that humans are responsible, either directly or indirectly; (2) those which see natural factors, primarily climatic change, as the primary cause; and (3) those which hold to a combination of anthropogenic and natural causes.

Before publication of Hastings and Turner's *The Changing Mile* in 1965, little attention was paid to the hypothesis that climatic shifts may have caused vegetation changes in southeastern Arizona since 1870. Previously, researchers believed that wildfire suppression or exclusion and overgrazing since Anglo settlement accounted for changes, at least in the rangelands. Hastings and Turner (1965), however, expressed doubt about this hypothesis because (1) they could find no evidence for frequent wildfires in the past and (2) they noted that the number of cattle in the region during the 1820s and 1830s was as great as in the 1880s and 1890s, and those cattle had little impact on the environment. To support their case for climatic change over a 120-year span, they claimed that their repeat ground photography of southern Arizona and northern Sonora, Mexico, clearly showed an upward displacement of plant ranges along a xeric-to-mesic gradient. They concluded that this change could only be the result of a trend toward greater aridity. The evidence examined here, however, contradicts their findings and conclusions, which are commonly quoted in the literature on vegetation change and arroyo cutting in the American Southwest and are often cited as evidence of recent climatic change in the region.

The principal objectives of this book are the following:

1. To determine what changes in the wild vegetation have actually occurred since 1870. Critical reviews are offered of all of the evidence for change—repeat ground and aerial photography, historic landscape descriptions, permanent plot studies, and the U.S. General Land Office surveyors' descriptions of the vegetation.
2. To determine which of the competing hypotheses for change (climate and land use) is best supported by the evidence.
3. To examine the historic land uses that have affected the native vegetation of southeastern Arizona since 1870, especially those which may have altered the vegetation at the turn of the century, and to establish whether these land uses were sufficiently intense to cause major directional changes in the vegetation.

There is little doubt that climatic oscillations since 1870 have resulted in short-term fluctuations in the vegetation (Martin and Turner 1977; Gehlbach 1981; Goldberg and Turner 1986). This book argues, however, that the most conclusive, long-term directional changes—such as the increase in woody plants in certain rangelands, the degradation of riparian wetlands, and the invasion and spread of exotics—have resulted from sustained and/or catastrophic human disturbances: overstocking and overgrazing of cattle, fuelwood cutting, wildfire suppression, agricultural clearing, construction of irrigation works, logging, railroad and road construction, and the intentional or accidental introduction of exotic plants.

Southeastern Arizona, one of the most ecologically diverse regions in the United States, is ideal for this study. Its vegetation and flora are better known than those of any region of comparable size in the western hemisphere (McGinnies 1981). Furthermore, there is a large body of literature on vegetation change in the region.

Few students of vegetation change in southeastern Arizona realize that nineteenth-century Arizonans exploited the wild vegetation at a pace equaling or surpassing that of many Third World countries, where destruction of native vegetation has led to major economic and social problems and to far-reaching ecological disequilibria. Moreover, although researchers frequently use early landscape descriptions and repeat ground photographs to reconstruct past landscapes and to identify vegetation changes, they usually do not identify or evaluate the historic land uses that may have caused the changes.

Assessing the extent to which plant communities have been modified by historic land uses is probably the most challenging problem in the study of vegetation change. In most cases, it requires years of diligent research through diverse and often obscure records just to ferret out bits of information on past land-use practices. This, combined with reconstructing and evaluating the effects of overlapping historic land uses on myriad ecosystems, makes studying the historic human impacts on vegetation a formidable task.

Incorporated in this study is a plea for students of vegetation change—in southeastern Arizona and in other world regions—to evaluate the historic and ecologic impact of traditional land uses on the evolution of the modern wild landscape and, in particular, to examine the links among different land uses and site-specific vegetation changes. This difficult research, essential to understanding the dynamics of vegetation and its recovery from human disturbance, will enable us to differentiate between the natural and anthropogenic factors of change.

Furthermore, this study, it is hoped, will serve as a model for future research on historic land-use and vegetation change not only in southeastern Arizona and in the rest of the American Southwest but also in other parts of the world. For the most part, our understanding of long-term land-use dynamics and vegetation change throughout the world is poor, and it is extremely hazardous to make sweeping generalizations about the causes and direction of vegetation changes based on a few short-term observations from a few sites (Ahlcrona 1988; Gritzner 1988).

No pretense is made that this book is a thorough evaluation of the hypotheses of change; rather, it is a land-use history of southeastern Arizona. Particularly emphasized are historic human impacts on southeastern Arizona's wild landscape—impacts that were just as dramatic and devastating a century ago as they are today.

Chapter 2 describes the physical setting and history of settlement of southeastern Arizona. Chapter 3 reviews previous studies of vegetation change, with particular emphasis on the evidence and hypotheses of change. Chapter 4 assesses observations of vegetation change from the

General Land Office surveyors' field notes and repeat ground and aerial photography, and examines the competing hypotheses of change (climate or land use) best supported by the evidence. Chapters 5 thru 9 define and analyze the major historic land uses that have affected the wild landscape since 1870. Chapter 10 identifies the patterns and causes of change.

2 Setting

PHYSIOGRAPHY

For this study's purposes, southeastern Arizona is defined as that portion of Arizona south of the Gila River and east of the Santa Cruz River and Highway 89 (see Figure 1.1). Classic, dry, basin and range country, it is characterized by a series of isolated, northwest-southeast trending, fault-block mountain ranges rising above broad, intervening valleys. In general, both the mountain ranges and the intervening valleys increase in elevation from west to east.

Four major northwest-southeast trending valleys dominate the region from west to east: the Santa Cruz, San Pedro, Sulphur Springs, and San Simon. Other important structural troughs of much lesser extent are the Aravaipa, San Rafael, and San Bernardino valleys. The lowest elevations in the study area are near Florence on the Gila River (1,490 feet); the highest elevations are in the Pinaleño and Chiricahua mountains (Mt. Graham at 10,713 feet and Chiricahua Peak at 9,796 feet). Considerable portions of the upland valleys east of the Santa Cruz Valley exceed 4,000 feet in elevation. With the exception of Willcox Playa and Whitewater Draw in the middle and southern parts of the Sulphur Springs Valley, all of the major drainage ways in the region eventually feed into the Gila River. The middle of the Sulphur Springs Valley is endorheic and drainage is into Willcox Playa (the only large basin of interior drainage in southeastern Arizona), while Whitewater Draw flows south into the Yaqui River of Mexico.

Toward the end of the nineteenth century the floors of the major valleys underwent rampant arroyo cutting (Cooke and Reeves 1976; Waters 1988; Betancourt 1990). Some researchers have equated this episode of arroyo cutting to local degradation of the vegetation cover by Anglo-American land uses; others have ascribed major weather anomalies, climatic change, diastrophism, or a combination of land-use and climatic factors. The problem of when, where, and why arroyo cutting took place in southeastern Arizona in the late nineteenth century is reviewed by Cooke and Reeves (1976).

CLIMATE

Southeastern Arizona's climate is semiarid.[1] Average annual rainfall ranges from nine to twenty-five inches; on some of the highest peaks, however, precipitation may approach thirty-five inches or more (see Figure 2.1) (Lowe 1964; Sawyer and Kinraide 1980:226). Generally, the amount of annual rainfall increases from northwest to southeast, although annual rainfall in the San Simon and Duncan valleys is less than ten inches (Sellers and Hill 1974). Rainfall tends to increase and temper-

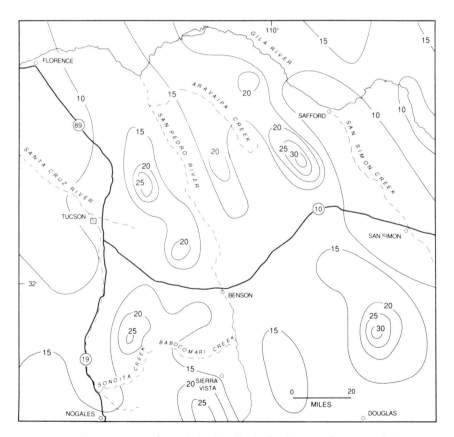

FIGURE 2.1. Average annual precipitation (in inches) in southeastern Arizona. *Source*: Sellers and Hill 1974.

atures tend to decline with increasing altitude. At the lowest elevations, average July temperatures are near 90°F. and average January temperatures are about 50°F.; at the highest elevations, average July temperatures are about 65°F. and average January temperatures are about 35°F.

The rainfall pattern in southeastern Arizona is bimodal, with summer and winter precipitation interrupted by spring and fall aridity. Summer precipitation during July, August, and September ranges from less than 50 percent of the annual total in the northwestern part of the area to more than 70 percent in the southeastern part; winter precipitation, which peaks in December and January, decreases from west to east. Precipitation characteristically varies from year to year. Winter precipitation, largely originating from storm centers in the westerlies, contrasts with the more intense and more localized summer rains associated with the influx of moist tropical air into southeastern Arizona.

VEGETATION

Southeastern Arizona's vegetation is extremely diverse, and the region includes most of Arizona's principal "life zones." In fact, the major vegetation differences can be ascribed to changes in elevation rather than in latitude and longitude. The "life zones," which are for the most part stratified altitudinally, include desertscrub (Sonoran desertscrub and Chihuahuan desertscrub), grassland (semidesert grassland and plains grassland), evergreen woodland (Madrean evergreen woodland), and, at the highest elevations, ponderosa pine and mixed-conifer forests (Rocky Mountain [Petran] and subalpine conifer forests) (see Figure 2.2).[2] Cutting across these zones and along the valley floors is riparian wetland.

Desertscrub

Sonoran and Chihuahuan desertscrub covers the floors of the major valleys, interdigitating with grassland. Sonoran desertscrub makes up 19 percent of the total vegetation cover of the region and dominates in the lower Santa Cruz, lower San Pedro, and Aravaipa valleys and on the terraces and fans along the Gila River (see Figure 2.3).[3] It merges in the east with semidesert grassland and, in some parts of the San Pedro and San Simon valleys, with Chihuahuan desertscrub. Chihuahuan desertscrub comprises 14 percent of the vegetation cover and dominates in the upper parts of the San Pedro, Sulphur Springs, and San Simon valleys, especially in Cochise County (see Figure 2.4).

Shreve (1942, 1951) recognized seven vegetation subdivisions of the

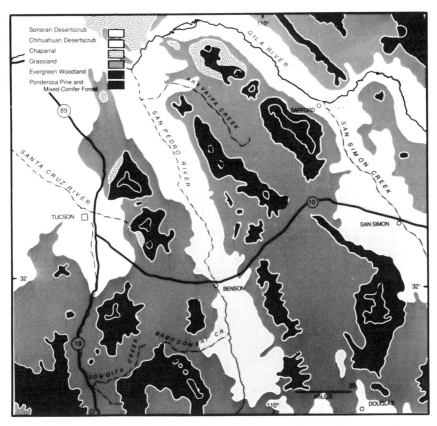

FIGURE 2.2. Vegetation cover of southeastern Arizona. *Source*: Brown and Lowe 1980.

Sonoran Desert; two make up the primary Sonoran desertscrub communities of southeastern Arizona—the lower Colorado River Valley and the Arizona Upland. The former, which occurs in the study area only in the lower Santa Cruz Valley and around Florence, is dominated by creosote bush (*Larrea tridentata*),[4] bursage (*Ambrosia* spp.), and saltbush (*Atriplex canescens*), interspersed with species of *Lycium* and *Prosopis*. Bunch grasses, such as big galleta (*Hilaria rigida*), and several native and introduced forbs also abound in this vegetation type. The Arizona Upland, often referred to as the paloverde-sahuaro community (*Cercidium-Carnegiea*), makes up the remaining Sonoran desertscrub in the region and has the appearance of a low woodland of leguminous trees with intervening spaces filled with shrubs and perennial succu-

FIGURE 2.3. Sonoran desertscrub (paloverde-sahuaro community, 3,100 feet, Santa Catalina Mountains). Photograph by D. A. Martinich.

lents. Among the most prominent species in the Arizona Upland are foothill paloverde (*Cercidium microphyllum*), sahuaro (*Carnegiea gigantea*), teddy bear cholla (*Opuntia bigelovii*), ocotillo (*Fouquieria splendens*), brittlebush (*Encelia farinosa*), ironwood (*Olneya tesota*), catclaw (*Acacia greggii*), bursage, and creosote bush. Scattered among the shrubs and trees are perennial and annual grasses.

Chihuahuan desertscrub, lying at elevations largely above 3,500 feet on mostly calcareous soils, is thought by several researchers to be invading former semidesert grasslands (Castetter 1956; Humphrey 1958; Hastings and Turner 1965; York and Dick-Peddie 1969). Early researchers believed, however, that Chihuahuan desertscrub is actually a grassland climax (Whitfield and Beutner 1938; Whitfield and Anderson 1938; Gardner 1951). The dominant plants in the Chihuahuan desertscrub are tarbush (*Flourensia cernua*), creosote bush, sandpaper bush (*Mortonia scabrella*), viscid acacia (*Acacia neovernicosa*), and yucca (*Yucca* spp.). Frequently scattered among these are crucifixion thorn (*Koeberlinia spinosa*), littleleaf sumac (*Rhus microphylla*), ocotillo, and honey mesquite (*Prosopis glandulosa*).

Grasslands

Semidesert grasslands and plains grasslands, the dominant cover in the study area, make up 46 percent of the total cover (see Figures 2.5 and 2.6). This figure seems much too high to me, however, and is based on the fact that most ecologists believe that before recent brush increases, the modern grass-shrublands were true grasslands. However, the semidesert grasslands might be better designated semidesert grass-shrublands because they are potentially perennial grass-shrub dominated landscapes located between the desertscrub below and the oak woodlands or plains grasslands above (Wright 1980). Isolated pockets of grassland also occur in Chihuahuan desertscrub. These pockets, referred to as *mogotes*, are dominated by tobosa (*Hilaria mutica*) and sacaton (*Sporobolus wrightii*).

The elevational limits of semidesert grassland are between 3,000 and 5,500 feet. Originally, grasses in this vegetation type were perennial bunchgrasses. Heavy grazing, however, has reduced bunchgrasses and increased low-sod grasses and annuals (Humphrey 1958; Brown

FIGURE 2.4. Chihuahuan desertscrub (tarbush–sandpaper bush–creosote bush community, 4,600 feet, Whetstone Mountains). Photograph by D. A. Martinich.

24   Historic Vegetation Change

FIGURE 2.5. Semidesert grassland (3,900 feet, near Pantano). Photograph by D. A. Martinich.

1982:124). Prominent grasses are the gramas (*Bouteloua* spp.), three-awns (*Aristida* spp.), tobosa, curly mesquite (*Hilaria belangeri*), cotton grass (*Trichachne californica*), and bush muhly (*Muhlenbergia porteri*). Interspersed among the grasses are species of *Opuntia, Yucca, Dasylirion, Agave,* and *Nolina*. Scrubby trees like mesquite (*Prosopis* spp.), acacia, and one-seed juniper (*Juniperus monosperma*) are also present. Mesquite is common at lower elevations; juniper is frequent at higher elevations. Much of this grassland has been invaded by Lehmann lovegrass (*Eragrostis lehmanniana*) and other exotic grasses and herbs. Increases in woody shrubs such as snakeweed (*Gutierrezia sarothrae*), burroweed (*Haplopappus tenuisectus*), and acacia—and in trees, such as mesquite and one-seed juniper—have greatly contributed to the "brushing up" of the grasslands. Common forbs are the filarees (*Erodium* spp.), lupines (*Lupinus* spp.), buckwheats (*Eriogonum* spp.), mallows (*Sphaeralcea* spp.), spiderlings (*Boerhaavia* spp.), white-mats (*Tidestromia* spp.), devil's claw (*Martynia* spp.), and the amaranths (*Amaranthus* spp.).

The plains grassland, also known as short-grass prairie, is restricted to elevations between 4,500 and 6,000 feet, especially in the Sonoita-Elgin area, the San Rafael Valley, and Allen Flat. This grassland, in which grasses form a mostly continuous cover, is dominated by such perennial grasses as the gramas (*Bouteloua* spp.), bluestems (*Andropogon* spp.), plains lovegrass (*Eragrostis intermedia*), threeawn (*Aristida longiseta*), galleta (*Hilaria jamesii*), and plains bristlegrass (*Setaria macrostachya*).

Evergreen Woodland

The evergreen woodland, made up largely of oak woodland, covers approximately 20 percent of the study area (see Figure 2.7). Dominated by woodlands of pure oak and oak-pinyon-juniper, it occurs at elevations between 4,000 and 7,000 feet. At lower elevations, the oak woodland is open; at higher elevations, particularly in moist environs, it can be dense. The most prevalent oaks are Emory oak (*Quercus emoryi*), Arizona white oak (*Q. arizonica*), and Mexican blue oak (*Q. oblongifolia*). Scattered among the oaks are alligator juniper (*Juniperus deppeana*), one-seed juniper, and Mexican pinyon (*Pinus cembroides*). At the upward margins of the evergreen woodlands, pines such as Apache pine (*Pinus engelmannii*) and Chihuahua pine (*Pinus leiophylla*) are found,

FIGURE 2.6. Plains grassland (4,900 feet, near Elgin). Photograph by D. A. Martinich.

as are manzanita (*Arctostaphylos* spp.) and madrone (*Arbutus arizonica*). At the lower margins, rosewood (*Vauquelinia californica*), one-seed juniper, and mesquite are frequently scattered among the oaks. The evergreen woodland has a large proportion of shrubby species at higher elevations that are progressively replaced at lower elevations by grasses, especially perennial bunchgrasses. Scattered about in the evergreen woodland, usually on thin or calcareous soils, are patches of chaparral dominated by manzanita (*Arctostaphylos pungens*), mountain mahogany (*Cercocarpus montanus*), buckbush (*Ceanothus greggii*), cliff rose (*Cowania mexicana*), silk tassel (*Garrya wrightii*), sumac (*Rhus* spp.), and rosewood.

Ponderosa Pine and Mixed-Conifer Forests

The ponderosa pine and mixed-conifer forests or Rocky Mountain (Petran) and subalpine conifer forests account for a little more than 1 percent of the total vegetation cover of southeastern Arizona and occupy elevations above 7,000 feet in the Santa Rita, Santa Catalina, Huachuca, Rincon, Chiricahua, Pinaleño, Santa Teresa, Winchester, and Galiuro

FIGURE 2.7. Evergreen woodland (extensive stand of evergreen oaks near Sheepshead, 5,800 feet, Dragoon Mountains). Photograph by D. A. Martinich.

FIGURE 2.8. Rocky Mountain (Petran) conifer forest (ponderosa pine, 8,400 feet, Rincon Mountains). Photograph by R. M. Turner.

mountains. The Rocky Mountain (Petran) conifer forest in southeastern Arizona is dominated by ponderosa pine (*Pinus ponderosa*), Douglas fir (*Pseudotsuga menziesii*), and white fir (*Abies concolor*), with scatterings of quaking aspen (*Populus tremuloides*), and Gambel oak (*Quercus gambelii*) (see Figure 2.8). The Rocky Mountain (Petran) subalpine conifer forest occupies only a few of the highest peaks in the Pinaleño, Santa Catalina, and Chiricahua mountains. These wet and cold forests average more than thirty-five inches of precipitation annually and are dominated by Engelmann spruce (*Picea engelmannii*), corkbark fir (*Abies lasiocarpa*), white fir, quaking aspen, and southwestern white pine (*Pinus strobiformis*).

Riparian Wetlands

Cutting across the insular "life zones" and lining the banks of the major rivers and streams are ribbons of galeria or riparian forests and wetlands. These wetlands, which make up much less than 1 percent of the total vegetation cover of southeastern Arizona, have been more affected by humans than any other major vegetation type, simply because

28 Historic Vegetation Change

FIGURE 2.9. Riparian wetland (cottonwood-willow galeria forest along Sonoita Creek near Patagonia, 3,900 feet). Photograph by D. A. Martinich.

of their proximity to water in a basically arid environment. Riparian forests are composed mostly of winter deciduous, broadleaf trees, although the dominant species generally vary greatly in type and composition with increasing elevation. The riparian forests that discontinuously line the banks of the major rivers of southeastern Arizona are predominantly of cottonwood (*Populus fremontii*) and willow (*Salix* spp.) with dense thickets of mesquite (mostly *Prosopis velutina* and *P. glandulosa*) (see Figure 2.9). Other important riparian trees are Arizona sycamore (*Platanus wrightii*), velvet ash (*Fraxinus pennsylvanica*), walnut (*Juglans major*), and saltcedar or tamarisk (*Tamarix chinensis*), the last an introduced tree that has now invaded nearly all of southeastern Arizona's major riparian habitats below 5,000 feet elevation.

SETTLEMENT BEFORE 1870

Humans have lived in southeastern Arizona for at least 12,000 years, but only the last 450 years of their occupancy have been recorded in writing (Haury et al. 1953; Haury et al. 1959). Consequently, very little is known about the impact of prehistoric humans on the evolution of the modern landscape. We cannot assume, however, that their impact was insignificant.

Numerous studies detail the effects of early and present-day Indians on the environment, and there is little doubt that they helped create the modern vegetational landscapes of much of the Americas. For example, intentional burning by Indians has created and/or maintained savanna and prairie vegetation in both North and South America (Sauer 1950, 1956, 1958; Stewart 1951, 1956; Hills 1969). Furthermore, researchers such as Cook (1949) have convincingly argued that long before Europeans arrived with plows and livestock, Indians were capable of causing drastic changes in the environment and of destroying physical resources.

Surely the prehistoric Indians of southeastern Arizona had the wherewithal to bring about major changes in the landscape through fire, agriculture (particularly with the construction of irrigation and water-retention works), hunting, fuelwood harvesting, and food gathering. Their impact on riparian habitats in particular may have been substantial (DiPeso 1951, 1953; Bahre 1977; Dobyns 1981; Waters 1988), especially after Old World crops and livestock were introduced. Waters (1988) notes that the prehistoric Indians may have increased the frequency of channel cutting and filling along the San Xavier reach of the Santa Cruz River.

No firm chronology has been presented of the prehistoric cultural occupation of the region, but both the archaeological record and early contact ethnographies indicate great differences in the technological capabilities and land-use practices of the prehistoric inhabitants. Some Indians lived in small bands and practiced hunting and gathering; others lived in large villages and depended on large-scale irrigated agriculture (Patrick 1903; Miller 1929; Judd 1931; Haury 1976). The Hohokam, for example, who inhabited central Arizona from about A.D. 900 to 1400, constructed 200 miles of irrigation canals in the Salt River Valley to irrigate extensive fields of maize and cotton (Masse 1981). At first Spanish contact, the successors of the Hohokam—the Pima and Sobaípuri—lived along the Santa Cruz, San Pedro, and Gila rivers, where they practiced ditch-irrigation agriculture and constructed retention-runoff structures for farming arroyo bottoms (Bolton 1919, 1936). These irrigation practices, along with clearing the land for crops, must have had a major effect in some riparian areas.

Prehistoric fuelwood harvesting probably had a negative effect on the trees and shrubs near the large settlements. In fact, because prehistoric Indians had no beasts of burden, the availability of nearby fuelwood must have influenced location of their settlements. Charles DiPeso (interview, Mar. 1976) believed that exhaustion of local fuelwood may have led to the abandonment of several large, prehistoric settlements in Arizona (see also Samuels and Betancourt 1982). Additionally, Indian hunting and burning may have contributed to major environmental changes. For example, P. S. Martin (1967) believes that the last of the Pleistocene megafauna were killed by prehistoric big-game hunters. Furthermore, Indian application of fire to the vegetation, especially the grasslands, may have accounted for the large expanses of brush-free grasslands in southern Arizona before Anglo-American settlement. Anthropogenic wildfire, however, probably added little to the high incidence of lightning fires.

In general, most prehistoric agricultural settlements were located near *ciénagas* or on the floodplains of the major perennial streams, where irrigated agriculture could be practiced. Prehistoric agricultural sites are most frequent on the floodplains and benches of the Santa Cruz River, San Pedro River, Gila River, Rillito Creek, Sonoita Creek, Aravaipa Creek, and Babocomari Creek near those sections with perennial flow.

Estimates of the aboriginal population of southeastern Arizona at initial Spanish contact vary, and the size of the prehistoric population may never be known. Sauer (1935:32), using early Spanish records, estimates that approximately 10,000 Indians lived in southeastern Arizona at contact. By the early 1700s, however, that population was greatly reduced by Old World diseases and Apache depredations. Furthermore, there is evidence that Old World diseases preceded the Spaniards into Arizona, killing hundreds, possibly thousands, of Indians before the 1540s (Dobyns 1963, 1978, 1981).

The first European to visit southeastern Arizona was probably Fray Marcos de Niza, who in 1539 traveled down the San Pedro River to seek the "Seven Cities of Cíbola," Zuni in present-day New Mexico (Bandelier 1890:147). There is some doubt, however, whether Fray Marcos traveled far into Arizona before returning south (Sauer 1932). Niza was followed in 1540 by Francisco Vásquez de Coronado, who journeyed the entire length of the San Pedro Valley on his way to Cíbola (Bolton 1949). Neither explorer gave Old World crops or livestock to the Indians, although large numbers of sheep and cattle accompanied Coronado on his expedition. Neither Niza nor Coronado expressed interest in the Indians and consequently left little information about them.

After Coronado's, apparently no other Spanish expeditions entered southeastern Arizona until 1692, when Father Eusebio Kino and a small military contingent under the command of Captain Juan Mateo Manje visited the Santa Cruz and San Pedro river valleys (Bolton 1919, 1936; Manje 1954). Kino established several *visitas* and churches in the region, then called Pimería Alta. He also introduced Old World livestock and crops to the Sobaípuri and Pima Indians. The Sobaípuri, however, who lived along the upper San Pedro River in the seventeenth century, had trade relations with Spanish colonists from the Rio Grande Valley between 1604 and the outbreak of the Pueblo Revolt in 1680 (Bolton 1952:428), so it is conceivable that European livestock, such as ducks, chickens, and rabbits, and some crop plants may have been obtained by the Sobaípuri and other Indians before 1692. According to DiPeso (1951:16), both peach and English walnut trees were grown by the Sobaípuri on Babocomari Creek long before Kino's arrival.

The Apaches did not move into southeastern Arizona until the 1680s or 1690s (DiPeso 1956; Aschmann 1970). Their depredations were so great in Pimería Alta that in 1764 Father Juan Nentvig, a Jesuit missionary, noted that 174 of 198 settlements (mines, forts, farms, and towns) in Sonora were depopulated or abandoned (Nentvig 1980:129). In Arizona, the Apaches forced the Sobaípuri to abandon the San Pedro Valley in 1762 (Nentvig 1980:73), and until the late 1870s they controlled the entire area south of the Gila River and east of the San Pedro River. In 1786 Viceroy Bernardo de Gálvez pacified the Apaches by bribing them with a fixed stipend of liquor and meat (Moorhead 1968; Sheridan 1986:18). This led to relative peace that lasted in Pimería Alta until Mexican independence in 1821.

The major Spanish settlements in southeastern Arizona during the Spanish occupation were the sparsely populated presidios of Tubac, Tucson, and Quiburi (Santa Cruz de Terrenate), and the mission settlements of Tumacácori, Guébavi, and San Xavier (del Bac) (Brinckerhoff 1967; Gerald 1968; Wagoner 1975; Officer 1987).

The introduction of European technology, systems of land tenure, mining, Old World food crops, and domesticated livestock during this Spanish occupation was to lead eventually to major changes in the landscape. However, because southeastern Arizona was on the margins of New Spain's northern frontier, Spanish settlement was sparse and concentrated in the upper Santa Cruz Valley. Livestock, especially cattle and horses, were abundant near the presidios and missions and may have substantially affected the adjacent rangelands, but there is no certain evidence that livestock numbers were particularly great at the time. For example, in 1804 Zúñiga reported that there were 3,500 cattle, 2,600 sheep, and 1,200 horses at Tucson, then the largest Spanish settlement in Arizona (McCarty 1976). Agricultural activities in conjunction with grazing near the presidios and missions probably significantly affected local riparian areas. Throughout the Spanish occupation, as noted previously, most of southeastern Arizona was in the hands of the Apaches, and although the Apaches relied heavily on horses in the nineteenth century, they seem to have had few in the eighteenth century.

With Mexican independence in 1821 the Apaches resumed raiding with devastating impact, especially in Sonora and Arizona (Weber 1982). Despite the Apaches, several large Mexican stock-raising grants were established from 1821 to 1827 (Mattison 1946; Officer 1987:149) (see Figure 2.10). All of the land grants, however, were abandoned by the 1830s or early 1840s because of Apache attacks (Mattison 1946). In 1900, many grants were given legality in Arizona by the U.S. Court of Private Land Claims and now represent the largest holdings of patent land in southeastern Arizona.

After 1846, the year war broke out between the United States and Mexico, several Anglo-American military and scientific expeditions and thousands of California immigrants traveled across southern Arizona.[5] Many of these travelers described in journals the abandoned haciendas of the Mexican land grants and large numbers of wild cattle, especially in the San Pedro and San Bernardino valleys (Bartlett 1854; Emory 1857; Cooke 1938; Goetzmann 1959).

The major towns in southeastern Arizona during the Mexican occupation were Tubac and Tucson, although there were small Mexican-Indian villages at San Xavier, Tumacácori, and Calabasas (Wagoner 1975). Except for the settlements of the upper Santa Cruz Valley and the Pima villages along the middle Gila River, the Apaches controlled all of southeastern Arizona east of the Santa Cruz Valley during the Mexican occupation. After only thirty-two years of occupation, the Mexican government in 1853 sold the United States all of present-day Arizona between the Gila River and the present international boundary. The so-called Gadsden Purchase was signed at Mexico City on December 30, 1853, and proclaimed on June 30, 1854.

The principal cultural impact bearing on the ecology of the region during the Mexican occupation was the introduction of large-scale cattle ranching in southeastern Arizona in the 1820s. During this period

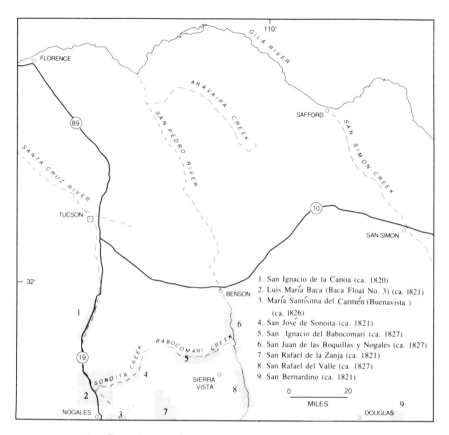

FIGURE 2.10. Confirmed Spanish and Mexican land grants in southeastern Arizona. *Source*: Wagoner 1975:158.

cattle raising may have weakened the native grass cover in the area adjacent to the present international boundary (the location of nearly all of the major Spanish and Mexican land grants in southeastern Arizona), especially in the bottomlands near water. There is, however, no certain evidence for this. Cooke and Reeves (1976) note, however, that erosion and stream entrenching, possibly due to agricultural disturbances and cattle grazing, had occurred along the Santa Cruz River near Tumacácori in the 1850s. In addition, Dobyns (1981) reports that the Pima Indians along the middle Gila River had large numbers of livestock during the Mexican occupation and the Apaches, large herds of horses. Irrigated agriculture near Tubac, Tumacácori, Calabasas, San Xavier, and Tucson must have significantly affected riparian habitats near those settlements. Small-scale mining was conducted in the Santa Rita and Patagonia mountains during the Mexican occupation, the mines depending on animal power for hoisting and crushing ores and on cordwood for making the charcoal used in smelting (Mowry 1864; Pumpelly 1918).

While several Anglo-American ranchers, miners, and farmers settled in southeastern Arizona shortly after the Gadsden Purchase, major Anglo settlement did not begin until a decade after the Civil War. Shortly after the Gadsden Purchase, however, the United States moved quickly to contain the Apaches, establishing Fort Buchanan on Sonoita Creek in 1857 and Fort Breckenridge (near the mouth of Aravaipa Creek) in 1859 (Serven 1965; Wagoner 1975). In 1862 Camp Bowie (Fort Bowie) was established in Apache Pass to protect stage and mail traffic across southeastern Arizona (Brandes 1960).

The U.S. Census of 1860 (U.S. Bureau of the Census 1864) listed a population of 1,635 for all of southeastern Arizona (see Table 2.1), but in 1864, a report on Arizona's population gave these statistics: Tucson (1,568), Mowry Mine (145), Apache Pass/Camp Bowie (74), San Pedro (6), Reventón and Calabasas (183), and San Xavier (112) (Wagoner 1970:41). Although these censuses did not usually include Indians, they made it clear that in the early 1860s southeastern Arizona was sparsely settled and that the bulk of the Anglo-Mexican population (approximately 95 percent) was concentrated in the upper and middle Santa Cruz Valley. After the Civil War, a few Anglo settlers moved into the Babocomari and Sonoita drainages to farm, and Camp Wallen and Camp Crittenden were established in 1866 and 1867, respectively, to protect settlers in those drainages (Brandes 1960).

Throughout the Spanish, Mexican, and early Anglo-American occupations to 1870, most of southeastern Arizona east of the Santa Cruz Valley was unsettled or under the control of the Apaches. Hence, probably only the Santa Cruz River Valley and the area along Sonoita Creek were much affected by the Spanish, Mexican, and early Anglo occupations before 1870, although cattle grazed the upper San Pedro, Babocomari, and San Bernardino valleys, especially along the California Trail. No doubt, Old World diseases killed off large numbers of Indians in southeastern Arizona, possibly even before initial Spanish contact, and Old World crops and livestock diffused throughout the region. The greatest environmental changes in southeastern Arizona in the past 450 years, however, appear to have come after 1870 with major Anglo-American settlement and economic development.

Table 2.1. Settlement and Population of Southeastern Arizona in 1860, 1870, and 1880

1860	1870	1880
Calabasas (17)	Adamsville (400)	Camp Bowie (184)
Casa Blanca (15)	Apache Pass or Camp Bowie (400)	Camp Grant (243)
Fort Buchanan (142)		Camp Thomas (112)
Lower Santa Cruz (63)	Calabasas (62)	Charleston (350)
Middle Santa Cruz (48)	Camp Grant (340)	Contention (150)
Patagonia Silver Mines (5)	Crittenden (215)	Dos Cabezas (126)
San Pedro (38)	Florence (218)	Florence (902)
San Xavier (35)	Goodwin (200)	Fort Lowell (227)
Sonorita (Sonoita) (13)	Rillito (32)	Harlowville (55)
Sonorita (Sonoita) Creek (100)	Saguano or Saginaw (71)	Maxey (145)
Sopori (13)	San Pedro (80)	Safford (173)
Tubac (310)	San Xavier (118)	San Jose (186)
Tucson (820)	Tubac (178)	Smithville (148)
Upper Santa Cruz (26)	Tucson (3,224)	Solomonville (175)
		Tombstone (973)
		Tucson (7,007)

SOURCE: U.S. Bureau of the Census (1864, 1872, 1883b).

SETTLEMENT AFTER 1870

Anglo-American settlement in southeastern Arizona expanded in the 1870s and 1880s, largely because of four interdependent, though distinct, events: the subjugation of the Apaches, the discovery of several important mining districts, the completion of the Southern Pacific Railroad from Tucson to El Paso, and a boom in cattle ranching. In 1870 nearly all non-Apache settlement was concentrated along the Santa Cruz and middle Gila rivers and at a few outliers such as Camp Bowie (Apache Pass), Crittenden, and San Pedro. Not until the campaigns of General George Crook between 1870 and 1872 were the Apaches finally subdued, although such renegade Apaches as Geronimo continued to raid throughout the region until 1886. In 1874 the White Mountain Indian Reservation (San Carlos) stretched south across the Gila River to Fort Grant, and the Chiricahua Indian Reservation (disbanded in 1876) occupied the region bounded roughly by the Dragoon Mountains on the west, Mexico on the south, New Mexico on the east, and Camp Bowie on the north (Wagoner 1970:145).

Establishment in the early 1870s of three major mining districts just north of the Gila River led to the first major intrusion of Anglo miners and settlers into the upper Gila River Valley and played a major role in the degradation of the riparian forests and mesquite *bosques* or woodlands of the middle and upper Gila River and its tributaries. These districts included the Silver King Mine, near Superior, discovered in 1872; the Globe Mine (Old Dominion), near Globe, founded in 1873; and the Longfellow Mine, near Clifton-Morenci, established in 1872 (McClintock 1916; Colquhoun 1924; Barnes 1935). The two biggest mining developments within the study area itself followed discovery of silver at Tombstone in 1878; as Tombstone's population boomed overnight, such stamp-mill towns as Charleston, Boston, Contention, and Fairbank were established, and Anglo-Americans developed the old Mexican copper mines at Bisbee in 1878 (Gird 1907; *Bisbee Review* Aug. 8, 1923; Spude 1979).

In 1881 the Southern Pacific Railroad was completed across southern Arizona (Greever 1957; Myrick 1975), facilitating growth of the mining industry and encouraging development of cattle ranching with the offering of low-cost transportation. The change in settlement and population between 1870 and 1880 is reflected in census figures (Table 2.1). For some reason, no population was given in the 1880 census for the thriving settlements of St. David, San Pedro, Calabasas, San Xavier, Tubac, Harshaw, and Fort Huachuca. While military camps were still important in 1880, the number of agricultural towns (mostly along the upper Gila River) and mining towns (mostly in the Tombstone District) increased significantly.

According to the *Annual Report of the Governor of Arizona* in 1879 (U.S. Congress, House 1879), all of southeastern Arizona had been prospected and mining operations were being conducted in most of the major mountain ranges. Between 1870 and 1900, mining brought many settlers into southeastern Arizona and spawned numerous small towns and mining camps, many of which are now abandoned. By 1882 major copper-mining activities were reported at Helvetia, Benson, Russelville, Bisbee, Clifton Morenci, Globe, Silver Bell, and Mammoth (*Arizona Daily Star* Nov. 12, 1882). At that time, cordwood was used to fuel the steam engines associated with mining and met all the heating and cooking fuel needs of the inhabitants (Bahre and Hutchinson 1985).

Sawmills in the ponderosa pine and Douglas fir forests of the Santa Rita Mountains supplied lumber to Tucson and nearby mining camps before 1870 (Matheny 1975:12). However, not until after settlement of the upper Gila Valley and the Tombstone and Bisbee mining booms were sawmills established in other mountain ranges of southeastern

Arizona having ponderosa pine and mixed-conifer forests: the Chiricahuas, Huachucas, Pinaleños, and Santa Catalinas.

From the time the Southern Pacific Railroad was completed until the early 1920s, a number of other railroads were built to meet the needs of mining and ranching (Myrick 1975). Among the more notable were the El Paso and Southwestern (Arizona and Southeastern); Arizona and New Mexico; Tucson and Nogales; Johnson, Dragoon, and Northern; Mascot and Western; and Arizona Southern.

Completion of the Southern Pacific also sparked a boom in cattle ranching, although Anglo-American ranchers had begun moving large herds of sheep and cattle into southeastern Arizona in the 1870s, especially after severe droughts in California (Haskett 1935; Morrisey 1950; Wagoner 1952). At that time, the grasses of southeastern Arizona were described as lush and abundant (ibid.). By 1885 so much investment capital had poured into southeastern Arizona's ranching industry that cattle numbers on the ranges exceeded all expectations. In 1891 more than 217,000 cattle were reported for Pima and Cochise counties alone—that number, according to some reports, was much too conservative (U.S. Congress, House 1893). Between 1891 and 1893, however, disaster struck: Drought and starvation led to the death of 50 to 70 percent of the cattle in southeastern Arizona (Wagoner 1952). Nevertheless, overstocking and overgrazing were to continue.

Like their Indian and Mexican predecessors, the early Anglo farmers depended on the few perennial streams and rivers for irrigation water. In the 1870s almost all irrigated farming in the region was concentrated along the Santa Cruz and middle Gila rivers, with the exception of some small outliers along Sonoita, Aravaipa, and Babocomari creeks and near the mouth of the San Simon River. In the early 1870s Mormon farmers established farms in the middle and upper Gila Valley between Bryce and Duncan, and in the San Pedro Valley between St. David and Hereford (McClintock 1921). There was some scattered pump irrigation and irrigation from artesian wells in southeastern Arizona after 1880, but until the 1940s most irrigated agriculture depended on surface flow from the perennial streams.

Early Anglo settlers and General Land Office administrators had a major impact on regional fire ecology by (1) initiating fire suppression policies and (2) overstocking and overgrazing the rangelands. Consequently, a major decline in wildfire frequency occurred throughout the region following Anglo settlement. This is documented by fire scars in tree-ring data from southeastern Arizona (Baisan 1988; Swetnam et al. 1989).

By 1900, output from the Tombstone mines had peaked and several mining towns were abandoned; the ranching industry had stabilized, but overstocking and overgrazing continued; agriculture was, for the most part, still concentrated along parts of the Santa Cruz, Gila, and San Pedro rivers; and there was growing interest in conserving range and forest resources. The copper mines at Bisbee, Globe, Superior, and Clifton-Morenci continued to prosper, and southeastern Arizona became one of the world's leading copper-producing regions. In addition, new mining towns such as Paradise, Courtland, and Gleeson were established. By that time, however, few mining operations still relied solely on fuelwood because local wood supplies were dwindling and cheaper fuels, such as coke, coal, and fuel oil, were becoming available. Nevertheless, cordwood remained the predominant fuel for household heating and cooking until the early 1940s (U.S. Bureau of the Census 1943:110, 114, 124).

The early twentieth century saw a major nationwide push to conserve range and forest resources, and in 1902, the first national forest reserves were established in southeastern Arizona in the Santa Rita, Chiricahua, Santa Catalina, and Pinaleño mountains (Lauver 1938; Baker et al. 1988; Allen 1989). Shortly afterward, other reserves were located in the Huachuca, Dragoon, and Whetstone mountains. In 1907 these reserves were renamed national forests, and by 1917 all of southeastern Arizona's national forest lands were consolidated under the Coronado National Forest. Soon after, a major effort was made to control and manage grazing, fuelwood cutting, and logging in the Coronado National Forest (ibid.).

Even in 1900, ranchers continued to overstock and overgraze the rangelands, although not at the levels of the 1880s and 1890s. For the most part, by 1900 the large ranches in southeastern Arizona had begun to acquire the smaller ranches (Wagoner 1952; Rodgers 1965:135), and this fact, along with the Stock Raising Act of 1916, which expanded the size of ranching homesteads, led to the death of the open range.

In 1906 the era of free use of the forage resources of the national forest reserves came to an end and the Forest Service placed new systems of grazing control in operation. Not until 1934, the year when the Taylor Grazing Act was enacted, however, was there an effort to stop injury to other public lands throughout the western United States and to stabilize the livestock industry dependent upon public range. In order to achieve these goals, grazing districts were established and grazing permits were issued. To keep their leases, the lessees had to improve the land. The Forest Service and other federal agencies charged with range and watershed protection introduced changes in the wild landscape by initiating land management practices designed to improve rangelands for cattle.

Between 1910 and 1940 most irrigation in southeastern Arizona still depended on surface flows from the Santa Cruz, San Pedro, and Gila rivers, although some acreage was irrigated from artesian wells, and by pumping in areas having high groundwater tables. In the 1910s and 1920s some homesteaders, to "prove up" their claims, attempted to dry-farm in the plains grasslands, but dry farming was rarely successful. In 1928, Coolidge Dam was completed on the Gila River to form San Carlos Reservoir; its construction not only led to agricultural expansion in the Florence and Coolidge areas but also created conditions for the widespread invasion of saltcedar or tamarisk along the Gila River and its tributaries (Harris 1966; Burkham 1970).

The major boom in irrigated agriculture in southeastern Arizona came in the early 1940s after development of more efficient, low-cost pumps. By the 1970s, hundreds of thousands of acres of grassland, riparian woodland, and desertscrub had been cleared for groundwater irrigation. Groundwater irrigation led not only to extensive land clearing but also to erosion and groundwater overdrafts. Groundwater overdrafts, in particular, had a major impact on riparian phreatophytes, killing extensive areas of mesquite and galeria forests. However, since the mid-1970s—primarily because of low agricultural commodity prices, increasing fuel costs, and the need to go to greater depths to reach groundwater—groundwater irrigation has declined rapidly and thousands of acres of cropland have been abandoned.

Mining activities today probably have less impact on the natural vegetation than in the past, mainly because the mines no longer depend on cordwood for fuel or local timber for construction. However, mining and smelting operations continue to cause water and air pollution, groundwater overdrafts, and some landscape destruction due to claim development, road building, and open-pit mining. Nevertheless, the area cleared for mining activities—mines, mills, smelters, tailing piles, ponds, waste dumps, and access roads—is minuscule compared with the total area of the region.

In 1970, 6 percent of the area in the Santa Cruz and San Pedro river basins was in irrigated agriculture, of which 33 percent (about 200,000 acres) was either idle or abandoned cropland, highways, roads, canals, ditches, and farmsteads (U.S. Department of Agriculture and Arizona

Water Commission 1977). In 1970, irrigated agriculture accounted for 88 percent of all groundwater use in those basins. The amount of natural recharge of the water tables in that year, however, was estimated at only one-third of agricultural use (ibid.).

Population increases in southeastern Arizona, especially since the 1940s, have led to expanded urban and rural development on privately owned lands and are threatening federal and state trust lands, which continually are being sold to the public to meet the pressures of development. One of the most pronounced changes in the landscape since 1950 has been an explosion in the number of rural subdivisions. For example, in 1970, 4 percent of the total area of the Santa Cruz and San Pedro basins was classified as urban or undergoing subdivision development (ibid.). Among Arizona's counties, Cochise County has the largest amount of private land (41 percent), and almost all of it is used for ranching, agriculture, or subdivision development (Hecht and Reeves 1981). Even cattle numbers in the county are declining because of tract development in former rangelands, and much private cattle and agricultural land is being purchased by people seeking rural retirement or investment opportunities.

In 1981 Santa Cruz County led the state in the percentage of land area in remote subdivisions, with 7.3 percent. Santa Cruz and Cochise counties were first and second in remote subdivision lot sales; most of these lots are bought as investments rather than as immediate homesites. The largest subdivision in southeastern Arizona is the 55,000-acre Rio Rico development, a few miles north of Nogales with 31,000 lots. The population capacity of subdivisions in Cochise County alone in 1974 was given as 254,000 (Hamernick and Brown 1975), yet the total population of the county in 1984 was only about 75,000. Tighter restrictions on developers in 1975 forced many major rural subdivisions into bankruptcy (Hecht and Reeves 1981), leaving thousands of acres with eroding, unused bulldozed roads and cleared surfaces (ibid.).

In 1970, 20 percent of all private land within the Santa Cruz and San Pedro basins had been converted to cropland, 2 percent to urban uses, and 9 percent to speculative land development (U.S. Department of Agriculture and Arizona Water Commission 1977). In those two basins, private land made up 32 percent of the total area (3,378,394 acres).

Many recent landscape changes in southeastern Arizona are due to increasing population pressure. Although several agricultural and mining towns have suffered major population declines since 1930 due either to the Great Depression or to the more recent major economic downturn in regional copper mining and irrigated agriculture, regional population growth has, for the most part, been steady and continuous. According to the 1980 census, the total population of southeastern Arizona was about 450,000, nearly 100 times the population reported in 1880 of 5,648 (U.S. Bureau of the Census 1883b). One city—Sierra Vista, which grew from a population of 3,121 in 1960 to 24,937 in 1980—now has more than 30,000. In 1980, cities of more than 3,000 in the study area included Tucson (330,537), Nogales (15,683), Douglas (13,058), Bisbee (7,154), Safford (7,010), San Manuel (5,443), Benson (4,190), Florence (3,391), Thatcher (3,374), and Willcox (3,243) (U.S. Bureau of the Census 1983).

Future population growth will lead to greater urbanization, which in turn will lead to intensified, multifaceted human impacts on vegetation, including air pollution and the effects of off-road vehicles. Of the study area covered by this book (about 13,690 square miles), approximately 30 percent is private land, 33 percent is state trust land, 18 percent is Forest Service land, 12 percent is Bureau of Land Management land, 5 percent is Indian reservation land, and 2 percent is military reservation, National Monument, Bureau of Reclamation, Bureau of Sport Fisheries and Wildlife, or state, county, and city park land.

3 Previous Studies of Vegetation Change

Numerous studies exist that deal with one or more aspects of vegetation change since 1870 in southeastern Arizona. For the most part, these studies have treated the following:

1. Changes in a particular vegetation type, usually the grasslands (Griffiths 1901, 1910; Thornber 1906, 1910; Leopold 1924; Parker and Martin 1952; Humphrey 1958, 1987; Martin 1975; Wright 1980) or riparian wetlands (Hastings 1959; Turner 1974; Brown et al. 1981; Dobyns 1978, 1981; Rea 1983; Hendrickson and Minckley 1984)
2. Changes in regional vegetation cover and/or changes in the vegetation along the international boundary (Humphrey 1956, 1987; Hastings 1959, 1963; Hastings and Turner 1965; Martin and Turner 1977; Bahre and Bradbury 1978; Gehlbach 1981)
3. Invasions and/or increases by native and exotic plants (Thornber 1906; Humphrey 1937, 1958; Glendening and Paulsen 1955; Tschirley and Martin 1961; Robinson 1965; Harris 1966; Cable 1971; Cable and Martin 1973; Burkham 1976; Horton 1977; Freeman 1979; Bock et al. 1986).

The main sources for identifying vegetation changes have been repeat ground photography and eighteenth- and nineteenth-century landscape descriptions. Also employed by researchers are nineteenth- and early-twentieth-century vegetation maps (Turner 1974), the U.S. General Land Office surveyors' field notes (Woodward 1972; Stoiber 1973), and permanent plots established near the turn of the century (Goldberg and Turner 1986). While the factors proposed to explain southeastern Arizona's vegetation changes vary widely, the most commonly mentioned are (1) climatic fluctuations or shifts, (2) cycles in rodent and lagomorph populations, (3) grazing by domestic livestock, (4) wildfire suppression, and (5) declining groundwater tables. The first four are the most frequently cited because they concern grassland, the vegetation most studied because of its historic importance to Arizona's cattle industry. The emphasis in these grassland studies has been on those changes detrimental to cattle production.

Physical and social scientists approach the causes of vegetation change in southeastern Arizona since 1870 from different viewpoints. Most physical scientists seldom identify human impacts, other than grazing and fire suppression, as having affected the environment, tending to consider human impacts insignificant compared with climatic change or grazing—except in riparian environments. For many, chronicling the succession of cultures that have occupied the area—Indian, Spanish, Mexican, and Anglo-American—and speculating on each culture's impact has been a sufficient explanation of the role of humans in the evolution of the wild landscape.

By way of contrast, social scientists (geographers, cultural ecologists, and environmental historians) emphasize humans as the primary cause of vegetation change; they tend to list human activities that may have affected the environment rather than to analyze the mechanisms of change or to examine the impact of each land-use practice historically and ecologically. In fact, no land uses such as grazing, woodcutting, or agricultural clearing have been analyzed, both historically and ecologically, and related to site-specific vegetation changes. What follows is a brief review of the most significant studies of vegetation change in southeastern Arizona since 1870. Emphasis is placed on evidence that has been used to identify the changes and the hypotheses proposed to explain them.

Early interest in vegetation change in southeastern Arizona grew out of two major environmental events near the end of the nineteenth century—the initiation of rampant arroyo cutting or stream entrenching along most of the major streams and rivers, and the increase in the rangelands of brush and scrubby trees. These two events, clearly recorded in historical writings and verified by studies of streambed stratigraphy, by repeat ground photography, and by General Land Office surveyors' field notes, had a negative impact on the livestock industry and irrigated agriculture, and were attributed by firsthand observers to the overgrazing and fire suppression that followed Anglo settlement (Toumey 1891a, 1891b; Griffiths 1901, 1910; Thornber 1907, 1910; Wooten 1916; Leopold 1924).

Thornber (1907, 1910), Griffiths (1910), Wooton (1916), and Leopold (1924) note that cattle, in particular, led to the increase in woody plants because they spread weeds, trampled and compacted the soil, and, in general, denuded the grass cover. Leopold (1924), pointing to the abundance of fire scars on trees throughout the region before the 1890s, attributes the increase in woody plants to a decline in wildfire frequency as a result of overgrazing. According to him and Griffiths (1910), woody plant increases were partly due to encouragement by early forest administrators to overgraze because this resulted in less fuel, and consequently in a reduced capacity for the rangelands to carry fire. To these administrators, overgrazing was preferable to having wildfires destroy fuelwood, timber, and pastures. Leopold (1924) further notes that scrub was encroaching on the grasslands, the evergreen woodlands, and the ponderosa pine forests; and, more important, that oaks and juniper were extending their ranges downslope into the grasslands. This contrasts, at least in the case of oaks, with the finding of Hastings and Turner (1965) that oaks were retreating upslope.

Alternative hypotheses have been advanced since these original observations were made to explain the increases in arroyo cutting and woody plants in southeastern Arizona as well as in other parts of the American Southwest. These hypotheses range from natural causes—climatic change and diastrophism—to various types of human disturbance. For example, some researchers believe that the degradation of the vegetation cover by overgrazing, in combination with woodcutting and agricultural clearing, led to arroyo cutting (Thornber 1910; Bailey 1935; Thornthwaite et al. 1942; Antevs 1952). Others believe that while overgrazing was the initiating factor, change to a drier climate (Bryan 1940; Judson 1952; Antevs 1962; Euler et al. 1979), to a more humid climate (Huntington 1914; Bryan 1922; Hall 1977), to either a drier or a more humid climate (Richardson 1945), or to a change in rainfall intensities (Leopold 1951a; Denevan 1967; Betancourt 1990) was also a major, if not the major, cause.

The initiation of gullying between 1850 and 1920 in valleys of southeastern Arizona, however, does not clearly correlate with periods of drought or heavy rainfall. Students of landscape change in southeastern Arizona have long held that roads, railroads, trails, and irrigation ditches exacerbated, if not caused, erosion and stream entrenching (Olmstead 1919; Leopold 1921; Wynne 1926). Cooke and Reeves (1976:94) conclude that discharges resulting from disturbances in the valley bottoms, such as the removal of vegetation by overgrazing, clearing for agriculture or woodcutting, destruction of beaver dams, and construction of roads, irrigation works, and railroad embankments, helped initiate historic arroyo cutting in southeastern Arizona. To support

Table 3.1. Relationships Between Disturbance and Stream Entrenching in Southeastern Arizona

Valley	Drainage Concentration Featured Associated with Entrenchment	Date of Origin of Feature	Date of Initial Entrenchment
San Simon	a. Irrigation ditch at Solomonville	1883	1883
	b. San Simon wagon road	by 1875	after 1885
	c. Railroad embankment	1884	after 1885
Aravaipa	a. Fort Grant wagon road	by 1875	after 1886
Whitewater	a. Levees	—	after 1884
	b. Cattle trails	—	after 1884
San Pedro	a. Canals, roads	locally before 1851	before 1851 in places
	b. Railroad embankment	—	—
Santa Cruz*	a. Greene's Canal	1910	1914
	b. Sam Hughes's Canal	after 1862	1883 or 1890
	c. San Xavier Canal	by 1851/1883	by 1871/c. 1883

SOURCE: Cooke and Reeves 1976:94.
*Betancourt (1990) notes that Sam Hughes's Canal was completed in 1888 and initial entrenching began in 1889. Also, according to Betancourt, the San Xavier Canal, which was first dug in 1849, was severely eroded by 1850.

their conclusions, they correlate the initiation of arroyo cutting in southeastern Arizona valleys with the occurrence of major bottomland disturbances (see Table 3.1).

Woody plant increases and the degradation of rangelands by cattle attract the most attention from researchers. Range scientists are particularly interested in the causes of woody plant increases and in ways to control the woody species and boost rangeland grazing capacity. Range improvement programs initiated by state or federal agencies have themselves led to extensive vegetation changes. Excellent reviews of the pertinent research on brush increases in the rangelands and range improvement in southeastern Arizona are found in Parker and Martin (1952), Humphrey (1958), Hastings and Turner (1965), and Wright (1980).

The riparian wetlands receive the second largest amount of attention from students of vegetation change. The evidence of changes abounds in the historic record and in repeat photography. Although riparian areas comprise a tiny percentage of the study area's total plant cover, they are often the most degraded simply because they have always been the focus of human settlement. While there has long been interest in riparian changes and arroyo cutting, ecologists have barely begun to examine the extent and causes of riparian vegetation degradation. Unlike changes in grasslands, changes in riparian wetlands are almost always attributed to human disturbance.

Henry Dobyns, an anthropologist, relying of eighteenth- and nineteenth-century landscape descriptions, examines the ways in which humans have historically degraded Sonoran Desert riparian environments in "Who Killed the Gila?" (1978) and *From Fire to Flood: Historic Human Destruction of Sonoran Riverine Oases* (1981). In *From Fire to Flood* he points out that ecologists have long ignored the effects of most prehistoric and historic human impacts on Arizona's environment. In particular, Dobyns feels the impact of Native Americans has been repeatedly underestimated by ecologists. He lists the following prehistoric, early historic, and nineteenth-century activities as having had a major impact on the evolution of southern Arizona's riparian wetlands or "Sonoran riverine oases": placer mining; excavation of "irrigation galleries" in alluvial deposits; digging flood diversion ditches; constructing poorly engineered irrigation canals and headgates; clear-cutting mesquite forests for charcoal making or agricultural exploitation on alluvium; upland forest clear-cutting near mine camps; clearing and traveling wagon roads; grazing and pasturing horses, donkeys, cattle, sheep, and goats with consequent trail trampling and stream bank caving; trapping beaver to extermination; suppressing Native American fire hunting; constructing railroad embankments; building bridges, thus constricting stream flows; and cultivating alluvial soils on steep slopes with moldboard plows.

Dobyns (1981:200) concludes that his research

> ... has shown that previous analyses of Sonoran Desert vegetational changes, erosion mechanisms, and related questions have been seriously flawed by overoimplification. Natural scientists concerned with these questions will hopefully recognize that human beings have played a more significant role in environmental change than previously thought.

Another work on riparian vegetation change in southern Arizona is Rea's *Once a River: Bird Life and Habitat Changes on the Middle Gila* (1983). The author, an ornithologist and ethnozoologist, discusses the negative impacts of Anglo settlement along the middle Gila River. He notes that the causes of riparian deterioration are (1) overgrazing of arid and adjacent semiarid uplands; (2) excessive woodcutting in watersheds and mesquite *bosques*; (3) overtrapping of beaver and loss of beaver dams; (4) gullying of stream banks and hillsides by the trampling of cattle and other Old World livestock; (5) cutting unprotected wagon roads and the stripping of these bare by horses, cattle, and wheels; (6) cultivating fragile desert and upland soils with methods suited to wet woodland agriculture; (7) developing water-control technologies unsuited to alluvial soils and short, intense precipitation; and (8) pumping groundwater far in excess of natural recharge (1983:3–4).

In "Ciénegas—Vanishing Climax Communities of the American Southwest" (1984), Hendrickson and Minckley, both biologists, attempt to integrate the natural and anthropogenic factors that have affected southeastern Arizona's riparian wetlands, especially the *ciénagas*. In evaluating changes in the *ciénagas*, they rely on nineteenth-century descriptions of the valley bottoms and wetlands. To assess the anthropogenic factors, they chronicle the succession of human cultures in southeastern Arizona and discuss the real and supposed impacts of each major culture on the riparian wetlands.

Studies of regional vegetation change spanning several different vegetation types are relatively recent. Leopold (1951b) made the first major study of vegetation change in the area, using repeat ground photography of the Gila River Basin with an average time between matched photographs of fifty years. He concludes, however, that the overall change in vegetation in his matched photographs is not sufficient for species counts to have any statistical significance.

In 1959 Hastings, a historian and meteorologist, published a paper on vegetation change and arroyo cutting in southeastern Arizona in which he relies on nineteenth-century landscape descriptions to measure changes in modern vegetation. He identifies four major changes: (1) a general decline in grassland, (2) degradation of marshes and springs, (3) rapid channel cutting, and (4) brush invasion. Hastings does not consider any human activity, other than cattle grazing, as a cause for regional changes, although he mentions the local impact of, for example, wild hay cutting and woodcutting. He concludes that overgrazing significantly affected the changes but is unclear about the role, if any, of climatic change.

In 1963 Hastings completed his doctoral dissertation, "Historical Changes in the Vegetation of a Desert Region." It contains ninety pairs of matched photographs of vegetation changes in the oak woodland, desert grassland, and desertscrub of southern Arizona and Sonora, Mexico.

In 1965 Hastings and Turner, a botanist, revised and expanded Hastings' dissertation into the now classic work on vegetation change in the Sonoran Desert, *The Changing Mile* (1965). They evaluate vegetation changes in ninety-seven pairs of matched photographs of southern Arizona and Sonora, Mexico (seventy matched photographs of thirty-six different sites in the study area), and conclude from changes in their matched photographs that since the 1890s (1) desertscrub and cactus communities have become sparser; (2) desert grasslands have receded, giving way to invasions by desertscrub, cacti, and mesquite; (3) former oak woodlands in the lower elevations are now dominated by mesquite; and (4) plant ranges have been displaced upward along a xeric-to-mesic gradient. They base their conclusions for the "upward retreat" of the evergreen woodlands on twenty-six matched photographs of fifteen different, mostly heavily disturbed sites. Seventeen photos show "severe attrition among the oak population" (1965:104). Of these, six are of four sites below 4,500 feet, and at three of those sites oaks have completely disappeared. Although a substantial part of *The Changing Mile* deals with "sequent occupance" and with assessing the impact of each major culture that has inhabited the region, there is no analysis of the impact of most historic land uses on vegetation. Hastings and Turner apply the four factors traditionally used to explain changes in the grasslands—cattle grazing, rodents and jackrabbits, wildfire, and climatic change—to explain changes in desertscrub, riparian wetlands, and evergreen woodlands. After considering each factor, they conclude that climatic change best accounts for regional vegetation changes since the 1880s and 1890s.

While Rodgers (1965), a geographer, is not specifically concerned with vegetation change in his study of settlement and landscape change in the upper San Pedro River Valley since 1870, he believes that a major consequence of early Anglo settlement was the rapid modification of local vegetation. He suspects that denuding the evergreen woodlands and mesquite *bosques* for fuelwood and the degradation of the grasslands by overgrazing in the late nineteenth century led to rampant arroyo cutting and the invasion of desertscrub into what was once a grass-dominated landscape. He points out that immediately after 1890, drastic changes occurred in the wild landscape of the upper San Pedro Valley, a landscape that had remained virtually unchanged for two centuries. He concludes that these changes did not correlate with periods of climatic change or any other natural factor, only with the initiation of major Anglo settlement.

The first land-use history of southeastern Arizona was undertaken by Bahre for the Research Ranch at Elgin in 1977. Like earlier researchers, he looks at "sequent occupance" in the upper San Pedro Valley and attempts to assess the impact on the environment of the Indians, Spaniards, Mexicans, and Anglo-Americans. In addition, he identifies and examines historically and ecologically several land-use patterns that have affected the region's vegetation: fuelwood cutting, mining, grazing, and exotic plant introductions.

In 1977 S. C. Martin, a range scientist, and Turner, perhaps realizing that short-term cyclic changes in vegetation were probably missed in matched photographs spaced more than fifty years apart, published six sets of photographs showing vegetation changes within the Sonoran Desert of Arizona and Sonora, Mexico. Each set of photographs comprises three to six separate photos taken on different dates to demonstrate both short- and long-term cyclic variations in the vegetation. Their photo sets demonstrate that plants such as foothills paloverde, sahuaro, and mesquite usually live the longest and are stable components of a community, while such plants as burroweed and bursage are short-lived and fluctuate markedly in number over short periods. Martin and Turner attribute changes in the vegetation primarily to natural forces, although they note that human-caused influences, such as grazing or removal of plants, can modify the impacts of natural factors. They also point out that changes wrought by nature can be as dramatic as those imposed by humans and that their pictures show that some vegetation changes are relatively permanent, while others are temporary.

In 1978 Bahre and Bradbury, both geographers, published "Vegetation Change Along the Arizona-Sonora Boundary." In this study they document, with repeat ground photographs of international boundary monument markers (nos. 98–124) and field sampling of the vegetation across the boundary, the historical and spatial differences in the desert grasslands, plains grasslands, and oak woodlands spanning the Arizona-Sonora boundary between the San Pedro River and the Pajarito Mountains. To assess both long- and short-term vegetation changes, they use their repeat photographs in conjunction with photos taken of the same markers by Charles Ames of the Coronado National Forest in 1969 (Ames 1977a, 1977b). Bahre and Bradbury observe probably the most significant aspect of vegetation change along the boundary—the contrast in vegetation cover across the international fence line. Space imagery (Gemini, Apollo, Skylab, and Landsat) and recent NASA Highflight aerial photography clearly show this contrast in vegetation communities that were homogeneous before completion of the fence line (see Figure 3.1).

Bahre and Bradbury conclude that this disparity has resulted from contrasting Mexican and Anglo-American systems of land tenure and land use, and simply cannot be explained by natural factors such as climatic change. They found that in the eastern section of the Arizona-Sonora borderlands (1) mesquite has increased throughout its range; (2) riparian forests have responded to changes in hydrologic regimes; (3) oaks have increased and decreased in different areas but their range has not changed; and (4) grass cover has increased overall since the 1890s. They reason that these changes were due largely, if not entirely, to historic land uses.

Gehlbach, a zoologist and ecologist, rephotographed 73 of the 205 monument markers along the international boundary between the Rio Grande and the Colorado River to assess regional vegetation changes over a ninety-year period in his study of the natural history of the U.S.-Mexican borderlands (1981). He wrote (1981:239) that "excluding cities, only 19 percent of my 73 photo pairs depict edificial influence, by contrast with 52 percent of the 64 matched pairs from the Hastings and Turner study [*The Changing Mile* (1965)]." Like Bahre and Bradbury (1978), Gehlbach finds no evidence for the upward retreat of vegetation types in the western borderlands. He particularly notes that most cyclic changes in vegetation are missed by historical-modern photo evidence spaced more than half a century apart and that frequently rephotographed sites confirm that cyclic ups and downs in vegetation are about two to five decades apart.

Gehlbach also points out that the major directional change in the borderlands is that they are brushier now than they were in 1890. He attributes this to continuing human impact in contrast with cyclic ties with nature. Like Hastings and Turner (1965) and Bahre and Bradbury (1978), he observes that native riparian deciduous trees in some areas are more abundant today than a century ago, probably because of changing hydrologic regimes. Gehlbach concludes that "Man's landscape tin-

FIGURE 3.1. Landsat image of the Arizona Sonora borderlands. Tonal contrasts across the international boundary between Arizona and Sonora demarcate vegetation differences. Minimal ground cover and a large amount of barren ground result in high soil albedo and light tonal signatures, while heavier ground cover results in low soil albedo and dark signatures. Courtesy of Arid Lands Studies, University of Arizona. The scale is approximately 1:1,100,000.

kering has been more influential than climate in creating unidirectional trends..." and that "... short-term climatic cycles not climatic change, have exacerbated man-made vegetative changes" (1981:239–241).

That changes in plant cover and density can fluctuate substantially over short periods because of climatic fluctuations has been further demonstrated by Goldberg and Turner (1986) in their study of vegetation change in several "protected" permanent plots at Tumamoc Hill, site of the former Carnegie Institution Desert Laboratory in Tucson. None of the plots has been grazed since 1907. Goldberg and Turner (1986:695) find that there "had been no consistent, directional changes in vegetation composition in the Tumamoc Hill plots between 1906 and 1978, despite large fluctuations in absolute cover and density of most species." This finding alone casts doubts on the role of natural factors in any directional vegetation change since the turn of the century.

In *90 Years and 535 Miles*, Humphrey (1987), a range scientist and biologist, rephotographed nearly all of the 205 international boundary monument markers between the Rio Grande and the Colorado River (46 in the study area) to determine vegetation changes and erosion since the markers were first photographed in the early 1890s. The repeat photography of the monument markers offers a more systematic means of assessing changes in the structure, composition, and distribution of regional vegetation types over ninety years than does the Hastings and Turner study (1965) because the markers are fairly evenly spaced along the boundary, the sites are less disturbed, and the boundary itself serves as an environmental transect passing through grassland, evergreen woodland, and desertscrub.

Humphrey concludes that since the early 1890s (1) many areas of Chihuahuan desertscrub that once supported grass or a grass-scrub mixture now support only scrub, (2) the grasslands have been invaded by woody xerophytes, and (3) evergreen woodland supports a much denser stand of trees and shrubs today than it did in the past. Like Bahre and Bradbury (1978) and Gehlbach (1981), he does not find evidence in the matched photographs of the boundary markers for changes in the areal extent of the major vegetation types in ninety years. Furthermore, Humphrey observes that almost all of the major vegetation changes have occurred in the heavily grazed eastern 40 percent of the boundary, where the perennial grasses or grass-scrub mixture that once prevailed over much of the area has been mostly replaced by pure or nearly pure scrub; meanwhile, the largely ungrazed 60 percent of the boundary west of the Tohono O'odham Indian Reservation has undergone no life-form or appreciable species changes since the 1890s. These findings further challenge the belief that climatic shifts are the major cause of vegetation change in the Arizona borderlands since 1870 because the most arid sections of the boundary have undergone little or no change, while the more mesic eastern sections, which are more densely settled and grazed, have undergone the greatest change. Humphrey deduces that changes in the vegetation were due to the interplay of climatic fluctuations, urban or rural development, grazing, and wildfire suppression (1987:429—430).

The literature on invasions and/or increases by native and exotic plants, such as burroweed, snakeweed, mesquite, tamarisk, and Lehmann lovegrass, is especially large and is reviewed in succeeding chapters. Although native plant increases appear to be the result of land degradation and wildfire suppression, some natives that have increased, such as burroweed and snakeweed, clearly respond to climatic fluctuations. The invasion of tamarisk, Lehmann lovegrass, and other exotics and increases of some natives, such as mesquite and acacia, are clearly related to human disturbance and land degradation. Probably the best synopsis of recent plant invasions in the American Southwest was done in 1966 by David Harris, an English geographer trained at Berkeley. Harris believes that woody plant increases resulted primarily from occupation of the Southwest by American settlers and the development of commercial livestock ranching, which led to increased seed dispersal, overgrazing, and the suppression of grass fires; these activities favored invasion of the grassland by woody plants. He notes that "short-term climatic fluctuations towards greater aridity have tended to accentuate rather than initiate the processes of invasion . . . and that these plant invasions illustrate how man may unintentionally bring about rapid and profound ecological changes in dry areas by the introduction of new systems of land use" (1966:408). Harris (1966:422) concludes:

> Indeed most arid areas have felt the impact, whether in the distant past or more recently, of the fundamental changes in land use that often accompany shifts in cultural dominance, often such as that from Indian hunter to American settler in the Southwest. The recency of this cultural mutation enables us to perceive some of its effects on plant distributions, but to explore through time the relationship between changes of culture and vegetation elsewhere in the dry lands, especially in those Old World deserts which have been occupied by man for millennia, is immeasurably more difficult.

Evidence of Change

Table 3.2 includes the major directional vegetation changes proposed in the studies reviewed above. The apparent lack of agreement among

Table 3.2. Proposed Vegetation Changes in Southeastern Arizona Since 1870

Researcher(s)	Expansion of Exotics	Decline in Grass	Increase of Woody Xerophytes	Decline in Riparian Wetlands	Increase in Native Riparian Trees in Certain Areas	Decline in Desertscrub and Cactus Communities	Extension of Oak and Juniper Ranges Downslope	Upward Displacement of Plant Ranges
Toumey (1891b)	X	X	X					
Griffiths (1901, 1910)	X	X	X					
Thornber (1906, 1910)	X	X	X					
Leopold (1924)		X	X				X	
Darrow (1944)	X	X	X					
Parker and Martin (1952)	X	X	X					
Humphrey (1958)	X	X	X					
Hastings (1959)	X	X	X	X				
Rodgers (1965)	X	X	X	X				
Hastings and Turner (1965)	X	X	X	X	X	X		X
Harris (1966)	X	X	X	X				
Turner (1974)	X	X	X	X				
Bahre and Bradbury (1978)	X		X	X	X			
Wright (1980)	X	X	X					
Gehlbach (1981)	X	X	X	X	X			
Dobyns (1981)				X				
Rea (1983)	X	X	X	X				
Hendrickson and Minckley (1984)	X	X	X	X				
Humphrey (1987)	X	X	X		X			

researchers about the types and extent of vegetation changes is due mostly to their different subsets of evidence and whether they examined changes in a particular vegetation type or over the entire region. No one reviewed all of the evidence of vegetation change in southeastern Arizona, even for a particular vegetation type; and more often than not, most researchers have relied on a particular subset of evidence, such as selected repeat photographs, to identify changes. Some investigators, however, have uncritically accepted the changes identified by others. Evidence for change has come primarily from two sources—historical landscape descriptions and repeat (historical-modern matched) ground photography. Most researchers seem to agree on the nature of the changes, but there are major exceptions.

Most of the changes in the rangelands (particularly in the grasslands) of southeastern Arizona were observed firsthand by several investigators (Toumey 1891a, 1891b; Griffiths 1901, 1910; Thornber 1906, 1910; Leopold 1924). Other researchers, such as Humphrey (1958), Hastings (1959, 1963), Hastings and Turner (1965), Rodgers (1965), Cooke and Reeves (1976), Dobyns (1981), and Henrickson and Minckley (1984), used early landscape descriptions from travel journals, reminiscences, and army and railroad surveyors' reports to assess changes in the modern landscape. Although historical landscape descriptions are often contradictory and colored by "good old days" fallacies, they clearly indicate that since the middle of the nineteenth century some riparian wetlands have been eliminated, grass has declined, and woody plants have increased in certain rangelands.

Hastings (1963), Hastings and Turner (1965), Bahre and Bradbury (1978), Gehlbach (1981), and Humphrey (1987) have used both historic landscape descriptions and repeat ground photography to identify changes. While the repeat photographs show vegetation changes, problems usually develop when researchers attempt to explain the cause of the changes or to use the changes themselves as evidence for climatic shifts or changes in regional vegetation distribution. Because presently available repeat photography covers such a small part of the region and because no land-use histories have been developed for any of the photographed sites, it is not known whether the vegetation changes in the photographs are a response to anthropogenic factors, to environmental factors, or to both.

Because Bahre and Bradbury (1978) were primarily studying changes in National Forest lands along a small segment of the boundary where grazing has been controlled since 1934, they identify an increase in grass since the drought of 1891–1893, when the earliest photographs were taken. In contrast with Hastings and Turner (1965), others, including Bahre and Bradbury (1978), Gehlbach (1981), and Humphrey (1987), find no change in oak ranges. In fact, Humphrey (1987) finds no change or an increase in junipers and oaks in all of the matched photographs of the twenty-one different monument markers in the evergreen woodlands along the boundary. In contrast, Hastings and Turner (1965) find a general die-off of oaks in the lower elevational ranges in their twenty-six photographs of fourteen different sites in the evergreen woodlands of southern Arizona and Sonora, Mexico. Hastings and Turner conclude that their evidence for oak decline, combined with woody plant increases in the grasslands, indicates an upward retreat of plant ranges since the 1880s.

In another interesting twist on the displacement of plant ranges, Leopold (1924) observes that oaks and juniper are moving downslope into the grasslands. It is well known that oak and juniper are increasing in southeastern Arizona, but whether this represents an invasion, an increase, or simply the reestablishment of these species in their former range is open to question. Furthermore, that desertscrub and cactus communities are becoming sparser, as Hastings and Turner (1965) observe, is not supported by succeeding studies by Turner and others (Martin and Turner 1977; Gehlbach 1981; Goldberg and Turner 1986). Finally, that native riparian deciduous trees, particularly willow and cottonwood, have established themselves in certain areas where they were absent a century ago is clearly demonstrated in repeat photography (Figures 4.3 and 4.12).

Proposed Causes of Change

Table 3.3 lists the causes (grouped into three categories) proposed by the various researchers of vegetation change in southeastern Arizona since 1870. Most of these investigators, however, have not conclusively evaluated their hypotheses for change and, at least in part, have agreed with the conclusions of their predecessors. While some hypotheses may explain changes in a particular vegetation type, they do not always explain changes in regional vegetation cover or vice versa. Although most of the researchers listed in Table 3.3 consider other causation factors, they were assigned the one they indicate best explains the changes. For example, whereas Hastings and Turner (1965) think grazing contributes to recent vegetation change, they propose that climatic change is most likely the primary cause of the woody plant increases in the rangelands and the supposed upward retreat of plant ranges in southern Arizona since the 1880s.

Table 3.3. Proposed Primary Cause of Vegetation Changes in Southeastern Arizona Since 1870

Human Activities	Human Activities and Climatic Change	Climatic Change
Toumey (1891a, 1891b)	Parker and Martin (1952)	Hastings and Turner (1965)
Griffiths (1901, 1910)	Hastings (1959)	Neilson (1986)
Thornber (1906, 1910)	Hendrickson and Minckley (1984)	
Leopold (1924)	Humphrey (1987)	
Darrow (1944)		
Humphrey (1958)		
Rodgers (1965)		
Harris (1966)		
Turner (1974)		
Bahre and Bradbury (1978)		
Wright (1980)		
Gehlbach (1981)		
Dobyns (1981)		
Rea (1983)		

Toumey (1891a, 1891b), Griffiths (1901, 1910), Thornber (1906, 1910), Leopold (1924), Darrow (1944), Parker and Martin (1952), Humphrey (1958), and Wright (1980) are interested primarily in changes in the rangelands, especially in the increase of xerophytes and the decline of grass. They conclude that these changes were due mainly to two interacting factors following Anglo settlement—fire suppression and overgrazing—although Darrow (1944) and Parker and Martin (1952) also note that climatic change and rodents and lagomorphs could have played a part in the increase of brush. Hastings (1959), in discussing changes in the biological environment of southeastern Arizona, considers grazing the major cause of vegetation change but is uncertain about the role of climate. He concludes (1959:66):

> ... when one looks at the facts of cattle population in the '80s, when one then looks at the incidence of flooding and cutting; when one sees that in 1882 (50,000 cattle), 7.08 mean inches of summer rainfall produced no unusual flood condition, whereas in 1886 (156,000 cattle), 4.63 inches did; when one looks at the damage accompanying 7.92 inches in 1890 (253,000 cattle), a very tempting conclusion presents itself.

In their examination of changes across a variety of major vegetation types, Rodgers (1965), Harris (1966), Bahre and Bradbury (1978), Gehlbach (1981), and Dobyns (1981) attribute the changes to human activities; Gehlbach (1981:230–241) points out that some cyclic changes were due to climatic and other natural fluctuations. Gehlbach (1981) and Harris (1966) conclude that while climatic fluctuations exacerbated recent alterations in vegetation cover, they did not initiate directional changes. Dobyns (1981), Turner (1974), Rea (1983), and Hendrickson and Minckley (1984) attribute changes in their riparian study areas in southeastern Arizona largely to human disturbance; Hendrickson and Minckley (1984) also consider the role of climatic shifts. The application of the climatic change hypothesis to riparian change, however, makes sense because many researchers believe that climatic shifts were the primary cause of arroyo cutting at the end of the nineteenth century. Finally, although Humphrey (1987) has taken a "holistic" approach to vegetation change along the border, attributing vegetation change to a combination of climatic and anthropogenic factors, he attributes the major changes to humans.

Chapter 4 assesses the nature and extent of vegetation changes since the 1870s, especially the directional ones. Emphasis is placed on the General Land Office surveyors' field notes and repeat ground and aerial photography. The chapter concludes with an analysis of the two hypotheses of change (climate and land use) that are best supported by the evidence.

4 Assessing Vegetation Change

To discern vegetation changes in southeastern Arizona since 1870, I examined repeat ground and aerial photography in conjunction with detailed land-use histories and the General Land Office surveyors' field notes to identify the types and directions of change in 100 randomly selected large-scale aerial photographs. These photographs were stratified proportionately, according to the percentage of the study area covered by each major vegetation type—Sonoran and Chihuahuan desertscrub (19 and 14 percent, respectively), grassland (46 percent), evergreen woodland (20 percent), ponderosa pine and mixed-conifer forest (1 percent).

GENERAL LAND OFFICE SURVEYORS' FIELD NOTES

The surveyors' field notes of the General Land Office (now the Bureau of Land Management) are available for all of the continental United States except for nineteen eastern and southern states. These notes, which contain descriptions of the vegetation along section and township lines, have been compiled since the initiation of the U.S. Public Land Survey (USPLS) in 1785. They have been used by many researchers to reconstruct the general characteristics and distribution of the vegetation as the first surveyors found it throughout the United States (Bourdo 1956). The original USPLS surveys of southeastern Arizona were conducted between 1872 and 1923, with a hiatus during the early 1880s when the Apaches were on the warpath.

Although Woodward (1972) and Stoiber (1973) relied on the field notes to evaluate local vegetation changes along the San Pedro River, the notes have not been used extensively to document vegetation changes elsewhere in Arizona. In New Mexico, however, Buffington and Herbel (1965) employed the surveyors' field notes to estimate changes in the grasslands of the Jornada Experimental Range.

Because of the obvious difficulties of mapping vegetation changes along every section line in southeastern Arizona from the surveyors' field notes, I selected 100 different section lines, one for each aerial photograph. To select the lines, I outlined the area covered by each aerial photograph onto the appropriate 7.5-minute and 15-minute U.S. Geological Survey topographic quadrangles. I then selected a section line within each outlined area. Descriptions of the vegetation along each section line were recorded from the surveyors' field notes at the offices of the Bureau of Land Management in Phoenix. Each section line was visited and field-checked between 1985 and 1987.

In addition, to obtain a better idea of the surveyors' biases and to gain a more complete picture of the nineteenth-century vegetation cover than is possible by reviewing scattered, individual section-line descriptions, I examined and field-checked the surveyors' descriptions of the vegetation of ten different townships. These townships were selected from twenty-six for which I had both repeat ground and aerial photography. They are T15S, R25E; T14S, R22E; T20S, R21E; T21S, R22E; T23S, R22E; T18S, R25E; T18S, R21E; T21S, R27E; T9S, R21E; and T30S, R32E.

For the most part, the original surveyors' descriptions of vegetation cover of southeastern Arizona are vague, incomplete, and often contradictory; as such, they are not reliable for measuring vegetation change. Thus, any map of the nineteenth-century vegetation cover of southern Arizona reconstructed wholly from the surveyors' field notes should be viewed skeptically. In many cases, if the vegetation descriptions are taken literally, it would be difficult to map, let alone interpret, the former vegetation cover. Nevertheless, the field notes are the only available on-the-spot information for most of Arizona's nineteenth-century vegetation.

EXAMPLES OF SECTION-LINE DESCRIPTIONS IN THE SURVEYORS' FIELD NOTES

The following descriptions represent the overall quality of the vegetation descriptions in the field notes and attest to the difficulties in using the notes to reconstruct vegetation cover and to measure vegetation change.

1. Western section line of S20, T13S, R16E. Original survey by T. F. White in 1873 (Book 779). (Describing a paloverde-sahuaro commu-

nity in Sonoran desertscrub near the Bellota Ranch east of Tucson; see Figure 4.1.)

"Land rolling. Soil 2nd rate. Some grass and scattering of trees."

2. Western section line of S28, T20S, R21E. Original survey by H. F. Duval in 1910 (Book 2279). (Describing Chihuahuan desertscrub near Fairbank.)

"North over rolling ground. Land rolling. Soil 2nd rate. No timber."

3. Western section line of S7, T12S, R14E. Surveyed by F. B. Jacobs in 1901 (Book 1527). (Describing Sonoran desertscrub in Cañada del Oro northwest of Tucson.)

"North across Cañada del Oro bottom through dense undergrowth of mesquite, paloverde, and catclaw. Land level and broken. Soil 1st and 3rd rate. Timber: paloverde and mesquite. Dense undergrowth of mesquite, paloverde, catclaw, and numerous cacti."

Description (3) is fairly informative, while it is difficult, if not impossible, to identify the vegetation cover in descriptions (1) and (2), let

FIGURE 4.1. Sonoran desertscrub along the western section line of S20, T13S, R16E, near Bellota Ranch, east of Tucson. Photograph by author.

alone use them to assess vegetation changes. Unfortunately, descriptions (1) and (2) are similar to the majority of the vegetation descriptions in the field notes. Moreover, the surveyors' accounts of the vegetation for the ten townships are just as vague and inconsistent as those for the individual section lines.

In most cases, so few different plant species are usually mentioned that one could mistakenly assume that any plant at present found along a section line and not recorded in the original survey is an invader. For example, sahuaro and mesquite were not recorded by the surveyors of the section lines near Bellota Ranch (W. Bd., S20, T13S, R16E, Book 779) (see Figure 4.1) and Redington (W. Bd., S14, T12S, R19E, and W. Bd., S28, T12S, R19E, Book 763), in areas that are presently luxuriant paloverde-sahuaro communities. Does one interpret this to mean that both sahuaro and mesquite, now abundant along these section lines, are recent invaders? Some researchers might say that the mesquite is, but that the sahuaro was simply overlooked.

Although the surveyors were required to indicate the names, diameters, and distances on-line to all trees with which the survey line intersected, and to note the different kinds of timber and undergrowth in order of dominance (U.S. Bureau of Land Management 1973), they did not always do so. Several did not consider mesquite, juniper, or paloverde to be timber, even though they often used them as bearing or marker trees. Moreover, individual surveyors described the same vegetation differently (e.g., W. and S. Township Bds. T18S, R21E; see books 889, 1509, 1533, and 3005). This was particularly the case when a section or township line had to be resurveyed. Finally, many surveyors only noted that there was no timber or that the grass was good or poor. In most cases, the surveyors seemed to be unfamiliar with the local vegetation and frequently did not know the common plants.

Nevertheless, I was able to conclude the following from my study of the field notes of southeastern Arizona:

1. The vegetation descriptions provide no evidence for the upslope retreat of oak woodland, the invasion of the grasslands by Chihuahuan or Sonoran desertscrub, or any change in the distribution of the major vegetation types of southeastern Arizona since 1870. For example, oaks were still found in 1987 along all of the twenty-three section lines in evergreen woodlands used in this study, except in areas that had been cleared (see Appendix A). Only along two section lines did oaks appear to be fewer today—the western section line of S32, T21S, R16E, just north of Patagonia (Book 956), and the western

section line of S27, T18S, R16E, northwest of the Empire Ranch near Highway 83 (Book 1857)—although both of the descriptions are vague and difficult to interpret. On the other hand, the field notes suggest an increase in oaks along two section lines—the western section line of S20, T17S, R31E, near Portal on the road to Paradise (books 1001 and 4024), and the western section line of S15, T22S, R16E, near Red Mountain south of Patagonia (books 919 and 2058).
2. The field notes indicate that since 1900, mesquite and acacia have increased mainly in the grasslands. For the most part, however, both species appear to have increased, not invaded, because they were recorded in the original surveys of most of the section lines where they occur today.
3. Grass conditions appear to have declined everywhere, although it is difficult to interpret what the surveyors meant by good, fine, or poor grass and good grazing. Because snakeweed and burroweed are rarely recorded in the surveyors' field notes, it is difficult to estimate how much these plants have increased since the 1880s. Nevertheless, they now dominate many overgrazed sites.

Probably the most significant finding in my historical-modern comparison of the vegetation along section lines is that the field notes offer no evidence for the premise that the distribution of major vegetation types in southeastern Arizona has changed since 1870. The conclusion that woody plants have increased while grass conditions have declined is well known to researchers of vegetation change in southern Arizona and has been unequivocally demonstrated in historic descriptions and repeat ground photography. For the most part, my findings are consistent with those of Woodward (1972), who, using the early descriptions of the vegetation in surveyors' field notes, finds no directional change in selected sections of Chihuahuan desertscrub in the Murray Springs and Tombstone areas since the late 1870s.

REPEAT GROUND PHOTOGRAPHY

Repeat ground photography, the practice of finding a historical photograph of a landscape, locating the original camera position, and taking a new photograph of the same scene, has been used extensively to study vegetation change in southern Arizona (Hastings and Turner 1965; Martin and Turner 1977; Bahre and Bradbury 1978; Gehlbach 1981; Rogers et al. 1984; Humphrey 1987). However, I could find long-term repeat ground photography for only 26 of the 100 aerial photographs selected.

Table 4.1. Sites, Geographic Coordinates, Negative Numbers in the Hastings-Turner Collection, Vegetation Types, and Elevations of the Ground Photographic Stations

Sites*	Lat. & Long.
Soldier Canyon	38°18'8"N., 110°44'41"W.
Rosemont Mine	31°50'N., 110°44'W.
Helvetia	31°52'N., 110°47'W.
Walnut Gulch	31°43'N., 110°9'W.
Tombstone	31°43'N., 110°5'W.
Millville-Charleston	31°38'N., 110°10'W.
Millville-Charleston	31°38'N., 110°10'W.
Contention (Fig. 4.2)	31°44'30"N., 110°11'20"W.
Pearce	31°54'3"N., 109°49'4"W.
San Rafael del Valle	31°36'N., 110°9'8"W.
Fairbank (Fig. 4.3)	31°43'20"N., 110°17'47"W.
St. David	31°52'5"N., 110°14'18"W.
"Hill of San Cayetano" (Fig. 4.4)	31°29'5"N., 110°49'44"W.
El Plomo Mine	31°36'41"N., 110°51'34"W.
Greaterville (Fig. 4.5)	31°45'47"N., 110°45'2"W.
Monkey Canyon	31°37'51"N., 110°41'52"W.
Sonoita Creek (Fig 4.6)	31°39'4"N., 110°42'9"W.
Fort Crittenden	31°39'50"N., 110°41'32"W.
Arizona-Pittsburg Mine (Fig. 4.7)	31°37'45"N., 110°51'44"W.
Aravaipa Canyon (Fig. 4.8)	32°36'7"N., 110°6'46"W.
West Stronghold Canyon	31°56'31"N., 109°56'56"W.
Commonwealth Mine-Pearce	31°58'54"N., 109°48'15"W.
Gleeson	31°44'5"N., 109°49'40"W.
Rosemont Mine (Fig. 4.9)	31°49'59"N., 110°43'50"W.
Pusch Ridge (Fig. 4.10)	32°23'40"N., 110°57'31"W.
Elephant Head	31°46'20"N., 110°52'12"W.
Davidson Creek	31°54'N., 110°40'W.
Red Rock Canyon	31°33'7"N., 110°43'1"W.
Santa Rita Experimental Range (Fig. 4.11)	31°51'28"N., 110°50'W.
Monument 111 (Fig. 4.12)	31°20'N., 110°36'W.

*Including the figure numbers of the photo sets reproduced here.
†The first number is for photographs in the Hastings and Turner Collection; the second number is for plates in *The Changing Mile* (Hastings and Turner 1965).

Neg. #s†	Vegetation Type	Elevation (feet)
4a (67)	Sonoran Desertscrub	2900
42	Evergreen Woodland	4800
44 (32)	Grassland	4300
49a (53)	Grassland	4000
54	Chihuahuan Desertscrub	4530
58 (50)	Grassland	4050
(51)	Grassland	4000
	Grassland	3900
123	Grassland	4440
134a	Chihuahuan Desertscrub	4050
150 (57)	Grassland	3800
152 (59)	Grassland	3700
158a (16)	Evergreen Woodland	4260
188 (1)	Evergreen Woodland	5050
200	Evergreen Woodland	5240
214a	Evergreen Woodland	4750
216a (6)	Evergreen Woodland	4750
217a (8)	Evergreen Woodland	4700
391a	Evergreen Woodland	5600
403a	Grassland	4380
420a	Grassland	4720
423b	Grassland	4650
424a	Chihuahuan Desertscrub	4950
1063	Evergreen Woodland	4750
236	Sonoran Desertscrub	2550
203a (38)	Grassland	3920
41 (37)	Grassland	4050
213a (28)	Grassland	4450
1175a	Grassland	4000
	Grassland	4639

Table 4.1 furnishes information on each of the thirty ground photo sites used. To evaluate both short- and long-term vegetation changes, I collected at least three separate ground photos taken on different dates for each site. The earliest photos were taken between 1882 and 1913, the middle photos between 1962 and 1986, and the most recent photos between 1983 and 1990. Except for one photographic set (Figure 4.2) from Bahre and Hutchinson (1985) and one set (Figure 4.12) from Bahre and Bradbury (1978), all of the early and middle photographs were furnished by Dr. Raymond Turner of the U.S. Geological Survey in Tucson from the Hastings and Turner Collection of Repeat Photography. Many of the matched photographs from that collection are reproduced in *The Changing Mile* (see Table 4.1). All of the most recent photos were taken with a 35-mm. camera. Because the first photographers used cameras with various focal lengths and because I was not always able to locate the original camera stations, some of the most recent photographs are slightly mismatched.

Since publication of *The Changing Mile* in 1965, Turner has continued to add to the Hastings and Turner Collection, which now includes nearly 600 matched photographs of sites in the study area. However, most of these photographs are of fewer than fifty different sites (many of the photos were taken in different directions from the same camera station or from closely contiguous camera stations). Consequently, the area covered by this long-term repeat ground photography is small. In addition, most of the early landscape photographs are of or include such cultural features as mines, stamp mills, townsites, reservoirs, corrals, farmsteads, ranches, military posts, and roads, and nearly all of the photo sites had been cleared, grazed, and/or cut over for fuelwood before and after the first photographs were taken.

Of the thirty matched ground photo sites used, nineteen were badly disturbed at one time, and none is free of the effects of clearing, grazing, and/or woodcutting. In addition, all of the sites in the original photographs had been disturbed by Anglo land uses before they were photographed. Eleven of the thirty photographic sets are reproduced here (Figures 4.2–4.12). The most obvious directional vegetational changes in the matched photo sets are a proliferation of mesquite, acacia, and other woody xerophytes, especially on degraded lands, and an increase in tree density and cover (particularly oak and juniper) in some evergreen woodlands in the Coronado National Forest. The latter change may be due to fire suppression and tighter controls on grazing and fuelwood cutting on National Forest lands since the 1930s. Another major change has been the appearance of cottonwoods and willows in riparian areas where they were absent in the 1880s and 1890s. This may have resulted from changes in hydrologic regimes caused by valley and upland disturbances, natural succession, or the reestablishment of these species in areas where they had been extirpated by woodcutting and grazing. In every case, however, oaks still occupy the same range, and there is no evidence for changes in the distribution of any major vegetation type. What changes have occurred appear to be random.

Based on this evidence, it can be argued that repeat photographs of southern Arizona show little more than revegetation of disturbed sites and/or short-term fluctuations in plant communities or natural plant life cycles, and certain directional changes, such as the increase of mesquite in the grasslands and the invasion of tamarisk in riparian habitats. Interpretation of vegetation changes in repeat photographs is complicated by lack of detailed land-use histories for the photographed sites and lack of information on the effects of most historic land uses on site-specific vegetation changes. Several plants identified by researchers as weedy invaders, such as juniper and desert willow, are natives that often appear to be reestablishing themselves in their former ranges after having been extirpated for fenceposts and/or fuelwood. Other plants, such as mesquite, acacia, burroweed, and snakeweed, favor areas of disturbance. Only in certain dynamic riparian environments are deciduous trees such as cottonwood and willow showing up for the first time in repeat photographs.

Based on my review of the vegetation changes documented in repeat ground photographs of southeastern Arizona, including the Arizona-Sonora boundary, I conclude that the changes in the photographs are probably due to the following factors:

1. Revegetation of once-cleared or cutover sites (e.g., mesquite in abandoned townsites) and/or recovery from past disturbances (e.g., increases in oak and juniper in the Coronado National Forest)
2. Rhythms in plant life cycles (e.g., cyclic ups and downs in burroweed) and/or short-lived cyclic responses of plants to both natural and man-made environmental fluctuations (e.g., increases in the cover of *Ambrosia deltoidea* after abundant winter rains)
3. Succession in highly dynamic environments, such as along unstabilized streams (e.g., recent cottonwood and willow establishment along parts of the San Pedro River), and/or changing human-caused hydrologic regimes (e.g., tamarisk invasion on the Gila floodplain below Coolidge Dam)

68 Historic Vegetation Change

FIGURE 4.2a. Contention, 1882.

FIGURE 4.2b. Contention, 1982.

Assessing Vegetation Change 69

FIGURE 4.2c. Contention, 1989.

FIGURE 4.2. Contention (1882[a], 1982[b], and 1989[c]), looking northeast down the San Pedro River from a terrace above the Contention Mill. Contention, one of Tombstone's major stamp mill sites, had a population of 150 in 1880 (U.S. Bureau of the Census 1883b). Contention is across the San Pedro River from Quiburi, the major settlement of the Sobaípuri Indians in the San Pedro Valley at initial Spanish contact and the location of the Spanish presidio of Terrenate in the 1770s (DiPeso 1953). Contention is also located in the middle of the old San Juan de las Boquillas y Nogales land grant (established 1827), which straddled the San Pedro River between present-day Charleston and St. David (Mattison 1946). In the original photograph taken by C. E. Watkins in 1882, large cottonwoods line the San Pedro River in the background. The original photograph also shows grassland along the river. In the 1982 and 1989 photographs the San Pedro River, which has meandered across part of the old townsite, is still lined by cottonwood, but the long-abandoned townsite and most of the former grasslands have been taken over by acacia and mesquite. The entire area has been heavily grazed since the 1880s. The site has recently been incorporated within the San Pedro Riparian National Conservation Area.

70   Historic Vegetation Change

FIGURE 4.3a. Fairbank, ca. 1890.

FIGURE 4.3b. Fairbank, 1962.

FIGURE 4.3c. Fairbank, 1990.

FIGURE 4.3. Fairbank (ca. 1890[a], 1962[b], and 1990[c]), looking east-southeast across the confluence of Babocomari Creek and the San Pedro River from a hill near Highway 82, about one mile west of Fairbank. The Mule Mountains are to the right and the Tombstone Hills are to the left. An irrigation canal (probably the old Ramon Escude irrigation ditch) follows the north bank of Babocomari Creek in the left foreground in the 1890 image. Hastings and Turner (1965) identified this canal as the main course of Babocomari Creek. The trees in the middle background mark the location of Fairbank, founded in 1882 as a railroad station for the New Mexico and Arizona Railroad, which ran via Fairbank from Benson to Nogales. Eventually, the El Paso and Southwestern Railroad and the Arizona and Southeastern Railroad also connected through Fairbank. In 1890 the tracks of the New Mexico and Arizona Railroad ran up the Babocomari Valley about one-half mile south of the camera station. The photo site is also in the middle of the San Juan de las Boquillas y Nogales land grant. Neither Babocomari Creek nor the San Pedro River appears to have a distinct channel in the original photograph, and grassland is extensive. In the 1962 and 1990 photographs, however, both streams have entrenched (the camera station for the recent photos is substantially northwest of the camera station in the original photo). Dike at left in the 1962 picture was constructed to divert Babocomari Creek. Mesquite, cottonwood, acacia, and willow now choke the valley bottom. The site has been included within the San Pedro Riparian National Conservation Area.

72   *Historic Vegetation Change*

FIGURE 4.4a. "Hill of San Cayetano," 1890.

FIGURE 4.4b. "Hill of San Cayetano," 1962.

FIGURE 4.4c. "Hill of San Cayetano," 1988. Photograph by D. R. Taylor.

FIGURE 4.4. "Hill of San Cayetano" (1890[a], 1962[b], and 1988[c]), from a station seven miles southwest of Patagonia and looking west toward what George Roskruge called the "Hill of San Cayetano," near Sonoita Creek (Hastings and Turner 1965:82–83). The Grosvenor Hills are at the right and the San Cayetano Mountains are at the left. In 1877 this site was less than one mile from the Davis Ranch and one-half mile south of the San Jose de Sonoita land grant (established 1821). The area along Sonoita Creek, which cut through the land grant from Calabasas to Patagonia, has been permanently settled since at least the 1860s. The New Mexico and Arizona Railroad, which was completed in 1882, followed Sonoita Creek to Calabasas. The site has been heavily grazed at least since 1860 and has been cut over in places for fuelwood. The site is also close to several major eighteenth- and nineteenth-century mining districts and mines (the Santa Rita, Salero, and Alto mines) (Pumpelly 1918:vol. 1, 197). Hastings and Turner (1965:82–83), noting that most of the trees in the original photograph were oaks, pointed out that they could not find a single living oak in the 1962 photograph. They did note, however, that "some relict Mexican blue oak can be found in sheltered spots nearby." They also noted a number of dead oak carcasses. Although mesquite have increased, healthy oaks are still abundant near the photo station, and several "dead oak carcasses" have stump-sprouted.

FIGURE 4.5a. Greaterville, 1896.

FIGURE 4.5b. Greaterville, 1962.

Assessing Vegetation Change 75

FIGURE 4.5c. Greaterville, 1990.

FIGURE 4.5. Greaterville (1896[a], 1962[b], and 1990[c]), looking southwest toward the abandoned townsite of Greaterville in the Santa Rita Mountains. The townsite is surrounded by the Coronado National Forest and is crossed by the road up Ophir (Hughes) Gulch. From 1875 to 1878 the gold placers around Greaterville were worked by 200 men (Wilson 1961:72). Large-scale mining of the gold placers continued off and on until 1948 (ibid.). According to Barnes (1935:190), the town, named Greater after an earlier settler, was called Santa Rita in 1873. Abandoned for many years, the original townsite is now much disturbed by visitors and cattle. Oak and mesquite are much denser, especially in the background, in the 1962 and 1990 images. This may be the result of fire suppression and tighter control over woodcutting and grazing on National Forest lands.

76   *Historic Vegetation Change*

FIGURE 4.6a. Sonoita Creek, ca. 1895.

FIGURE 4.6b. Sonoita Creek, 1962.

FIGURE 4.6c. Sonoita Creek, 1988.

FIGURE 4.6. Sonoita Creek (ca. 1895[a], 1962[b], and 1988[c]), looking north-northwest from a station about three miles southwest of Sonoita. In the midground (a and b), Sonoita Creek winds under a trestle of the New Mexico and Arizona Railroad (established 1882). The site is within the long-abandoned Camp Crittenden Military Reservation (established 1867). The old site of Fort Buchanan, founded in 1857, is at the mouth of Hog Canyon in the middle background of the 1895 photograph. The camera station is one-half mile from the Rail X Ranch, established in the 1880s. The trees are mostly oaks. In the 1962 and 1988 images, trees and shrubs seem to be much denser. The first photo was taken just two years after the 1891–1893 drought. Houses are now scattered throughout the scene.

FIGURE 4.7a. Arizona-Pittsburg Mine, 1909.

FIGURE 4.7b. Arizona-Pittsburg Mine, 1968.

Assessing Vegetation Change 79

FIGURE 4.7c. Arizona-Pittsburg Mine, 1989.

FIGURE 4.7. Arizona-Pittsburg Mine (1909[a], 1968[b], and 1989[c]), looking northeast toward the main shaft of the Arizona-Pittsburg Mine and located two miles north-northeast of Alto in the Santa Rita Mountains. Few, if any, changes have occurred in the oaks or the rest of the landscape since 1909. An 1890 plat map of the area shows a number of old smelters and charcoal-burning pits near the old Salero Hacienda, about three miles southwest of the camera station. The entire area between Josephine Canyon in the north and Squaw Peak in the south has been mined and heavily grazed since the mid-eighteenth century. The mine is now within the Coronado National Forest.

80 Historic Vegetation Change

FIGURE 4.8a. Aravaipa Canyon, ca. 1910.

FIGURE 4.8b. Aravaipa Canyon, 1968.

FIGURE 4.8c. Aravaipa Canyon, 1988.

FIGURE 4.8. Aravaipa Canyon (ca. 1910[a], 1968[b], and 1988[c]), looking northwest toward the Pinaleño (Graham) Mountains across the head of Aravaipa Canyon, near Hooker's Sierra Bonita Ranch. H. C. Hooker, who brought large numbers of cattle into the region in the early 1870s, was the first Anglo rancher in the Sulphur Springs Valley and Aravaipa Canyon. In the midground of the 1910 image is the old Globe-Willcox road. The camera station is 2.5 miles from Hooker Ciénega and about 6 miles from the Sierra Bonita Ranch. All of the photographs show evidence of heavy grazing. The 1968 and 1988 images show erosion along the old roadway and the establishment of burroweed, snakeweed, acacia, and mesquite.

82 *Historic Vegetation Change*

FIGURE 4.9a. Rosemont Mine, ca. 1906.

FIGURE 4.9b. Rosemont Mine, 1982.

FIGURE 4.9c. Rosemont Mine, 1990. Photograph by D. A. Martinich.

FIGURE 4.9. Rosemont Mine (ca. 1906[a], 1982[b], and 1990[c]), looking west toward the Rosemont Mine and stamp mill in Barrel Canyon (Santa Rita Mountains). In the 1906 picture few, if any, junipers are readily identifiable; in the 1982 and 1990 images junipers have increased dramatically and oaks have remained constant if not increased. Some stumps (juniper?) are visible in the 1906 photograph. Juniper was much in demand for fuelwood, fenceposts, and mine timbers at the turn of the century and, unlike oak, was often killed by cutting (Bahre and Hutchinson 1985). In the 1906 photo, oaks in the immediate vicinity of the mine seem to have been left alone, while those on the midground slopes between the houses and the stamp mill appear to have been cut over. The earliest mining claims adjacent to the Rosemont Mine, according to BLM records, were filed in 1898. In a plat map of the area done in 1898 there were many claims, several windmills, and Vail's corral in the vicinity. The road up Barrel Canyon past the Rosemont Mine was also shown on the 1898 plat map. The Rosemont site was grazed, cut over, and mined several years before the first photograph of the mine was taken. The area is still grazed today.

84   Historic Vegetation Change

FIGURE 4.10a. Pusch Ridge, ca. 1912.

FIGURE 4.10b. Pusch Ridge, 1962.

FIGURE 4.10c. Pusch Ridge, 1990. Photograph by D. A. Martinich.

FIGURE 4.10. Pusch Ridge (ca. 1912[a], 1962[b], and 1990[c]), looking east-southeast toward Pusch Ridge from the east side of Highway 89 near the former Pusch Ranch (Steam Pump Ranch), about five miles north of Tucson on Highway 89. In 1902, according to early plat maps, roads, telegraph lines, and fences crisscrossed the section in which the photo site is located. The site has been grazed by cattle since the mid-nineteenth century. In the 1962 image the paloverde tree that was in the foreground in the original image has died, but the rest of the vegetation seems the same. In the 1912 image burroweed occupied the range in the foreground, as it does today. Housing developments now surround the site, and a drainage canal can be seen in the immediate foreground of the 1990 image.

86   Historic Vegetation Change

FIGURE 4.11a. Santa Rita Experimental Range, 1902.

FIGURE 4.11b. Santa Rita Experimental Range, 1986.

FIGURE 4.11C. Santa Rita Experimental Range, 1989. Photograph by D. Lippert.

FIGURE 4.11. Santa Rita Experimental Range (1902[a], 1986[b], and 1989[c]), looking east toward the Santa Rita Mountains from the top of Huerfano Butte in the Santa Rita Experimental Range just southeast of Tucson. The Santa Rita National Forest Reserve was established in 1902, and the Experimental Range was not set aside until a few years later (Barnes 1935:390). Parts of the Experimental Range have been continuously grazed since the 1850s, if not before (Pumpelley 1918). The area was also a major source of wild hay in the 1880s and 1890s (Bahre 1987; see Figure 9.3 in this volume). The 1902 image shows an overgrazed rangeland that was still fairly brush-free. In the 1986 and 1989 images, mesquite and acacia have increased dramatically. The mining area in the middle background is Helvetia.

88 Historic Vegetation Change

FIGURE 4.12a. Monument 111, 1893.

FIGURE 4.12b. Monument 111, 1969.

FIGURE 4.12c. Monument 111, 1983. Photograph by R. R. Humphrey.

FIGURE 4.12. Monument 111 (1893[a], 1969[b], and 1983[c]), on the U.S.-Mexico international boundary in the middle of the San Rafael Valley on the old San Rafael de la Zanja land grant (established 1821). This sequence of photographs, which looks east toward the headwaters of the Santa Cruz River, shows a striking change in the riparian vegetation along the stream course since 1893. Although the Santa Cruz River was reported to be perennial and rimmed by galeria forest in the 1880s, no trees can be seen along the river in the 1893 scene. In the photographs taken in 1969 and 1983, cottonwood and black willow line the stream banks. The old adobe building (in the 1893 photo) that Emory visited during his boundary survey in 1855 has crumbled into ruin, and only sacaton marks its former location. The grass cover has improved since the 1893 view of an almost completely denuded prairie. The 1893 photo was taken near the end of one of the worst droughts in Arizona's history. The area is still heavily grazed on both sides of the international boundary.

4. Long-term land-use impacts such as grazing (e.g., mesquite increase in the grasslands), wildfire suppression (e.g., change in ponderosa pine forest structure), groundwater withdrawal (e.g., death of phreatophytes on river benches), and clearing for settlement.

Three directional vegetation changes are clearly documented in repeat ground photography: an increase in xerophytic trees and shrubs in various localities, especially in grazed rangelands; the introduction of exotic plants; and the continual extirpation of native plant cover for settlement. There is no evidence in the repeat photography of southeastern Arizona for directional changes in the distribution of any major vegetation type; if anything, the bulk of the photographic evidence points to no change in native plant ranges except possibly in areas of continual disturbance.

REPEAT AERIAL PHOTOGRAPHY

For assessing changes in vegetation distribution, repeat aerial photography is superior to repeat ground photography, primarily because vertical aerial photographs are taken perpendicular to the earth's surface and thus are more useful for planimetric measurements. More important, because of their contiguous cover, it is possible to select truly random sites for analysis. Unlike ground photographs, with their narrow, oblique fields of view, vertical aerial photographs show the distribution and patterning of vegetation over large areas. There are, however, two disadvantages in using repeat aerial photography to assess vegetation changes in southeastern Arizona: the time span between the matched-photo coverage is short, because the first vertical aerial photographs were not taken until the mid-1930s, and the resolution of the early aerial photos is often poor. Nevertheless, repeat aerial photographs provide the most foolproof method for measuring changes in vegetation distributions.

In this study I used Soil Conservation Service (SCS) photographs made between 1935 and 1937 and National High Altitude Photography (NHAP) photographs acquired in 1983 and 1984. In the early 1930s the SCS commissioned the first aerial surveys of southeastern Arizona—the Gila River Indian Reservation Survey, completed in 1935, and the Pima-Papago Indian Reservation Survey, completed in 1937. These surveys, which cover all of the study area except for the southeasternmost corner—the Chiricahua Mountains and the San Bernardino Valley—consist of a series of overlapping panchromatic vertical photographs having a scale of 1:31,680. For the most recent coverage of the entire area I used the color-infrared NHAP imagery flown between June 24 and July 5, 1983, and June 7 and June 9, 1984, at a scale of approximately 1:58,000. Because of its high resolution, the NHAP imagery can be magnified to match the scale of the SCS imagery.

As noted at the beginning of this chapter, I selected 100 nonoverlapping SCS aerial photographs stratified according to the percentage of cover in each major vegetation type in the study area (forty-six images in grassland, twenty in evergreen woodland, nineteen in Sonoran desertscrub, fourteen in Chihuahuan desertscrub, and one in ponderosa pine and mixed-conifer forest). Riparian vegetation was not included as a separate category because it amounts to less than 1 percent of the total vegetation cover and is found in all of the major vegetation types.

Because I was unable to examine each SCS image of the study area at the National Archives, I used two criteria for selecting the 100 images. Believing that it would be useful to have both matching ground and aerial photographs of my sites, I obtained from Raymond Turner the geographic coordinates for nearly 600 camera stations for which he has matched ground photographs in the study area. As I previously pointed out, however, only twenty-six SCS aerial photos were needed to include all of these camera stations. Once these twenty-six images were grouped according to vegetation type, the remaining seventy-four aerial photos were randomly selected by locating points within each major vegetation type on Brown and Lowe's map of Arizona's vegetation (1980). (See Figure 2.2.)

The relatively small scale of the repeat aerial photography limited my observations to (1) comparing the number of trees and large woody shrubs along randomly selected transects on the matched photos and (2) evaluating changes in vegetation patterns by overlapping the NHAP images onto the SCS images, using a Zoom Transfer Stereoscope (see Figures 4.13 and 4.14). Because several SCS photographs were not dominated by one vegetation type, I had to compensate by using two transects per photograph to measure changes in tree and woody shrub numbers in each vegetation type rather than one transect. The transects

FIGURE 4.13. Matching (a) SCS (1936) and (b) NHAP (1984) aerial photographs of oak-grassland near Nogales. The scale is approximately 1:10,000.

FIGURE 4.14. Matching (a) SCS (1936) and (b) NHAP (1983) aerial photographs of oak-grassland near Sonoita. The scale is approximately 1:10,000.

92　Historic Vegetation Change

FIGURE 4.13a.

Assessing Vegetation Change 93

FIGURE 4.13b.

94 Historic Vegetation Change

FIGURE 4.14a.

FIGURE 4.14b.

varied from one-quarter mile to slightly more than three miles in length. The transects were then transferred to the NHAP imagery, and tree and woody shrub counts were conducted along each of the matching transects. I used a Zoom Transfer Stereoscope both to magnify the photographs and to match the scales of the SCS and NHAP imagery.

I sampled 111 transects instead of 100 because I was unable to assign some transects to one or another vegetation type due to local anomalies. Nevertheless, the numbers and groupings of transects still represent the percentages of cover in each major vegetation type. Once the transects were completed and grouped according to vegetation type, I compiled histograms showing absolute changes in tree and woody shrub counts per transect and percentage changes in counts per transect between the SCS and NHAP imagery for the entire sample and for each major vegetation type (see Appendix B, and Figures 4.15 and 4.16). In addition, a logarithmic transformation was performed on the data. This transformation compensates for the large differences in the density of trees and woody shrubs between different communities.

To determine if the two temporal sets of data were significantly different, I used the nonparametric Wilcoxon Signed-Rank Test with logarithmically transformed data to compare the tree and woody shrub counts per transect in the matched imagery (Sokal and Rohlf 1987). The

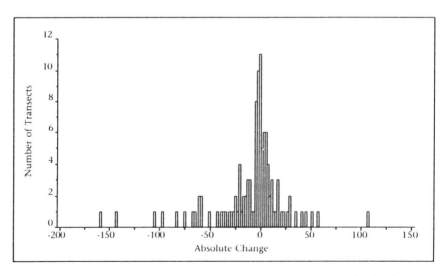

FIGURE 4.15. Histogram of absolute change in tree and large woody shrub numbers by transect between 1935 and 1984.

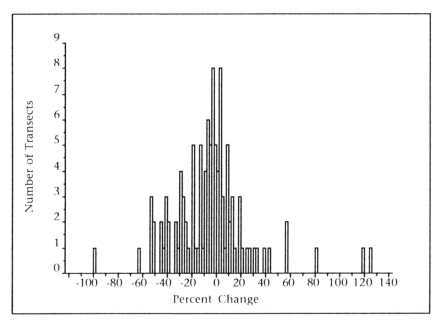

FIGURE 4.16. Histogram of percentage of change in tree and large woody shrub numbers by transect between 1935 and 1984.

test was selected because the samples are small and sometimes skewed, and the same transect is studied at two different periods of time. The test statistics were expressed as a Z-score corrected for ties. A significance level of .05 was used, and significance was tested by comparing Z-scores with standardized Z-scores for a normal curve. In addition, I used the same test to compare transects according to vegetation type (Table 4.2; see also Appendix C). Ponderosa pine and mixed-conifer forest (only one transect) was not considered in this comparison.

In the first analysis, the Z-score for the total sample was statistically significant (Table 4.2). When the significance was considered by individual vegetation type, however, only Chihuahuan desertscrub was significantly different. Therefore, to consider the contribution of individual vegetation types to vegetation change within the study area, Z-scores were computed for combinations of vegetation types, for example, the total sample minus Chihuahuan desertscrub. When the Chihuahuan desertscrub data set was removed from the total sample, the Z-score was still significantly different for the aggregate of the remaining vegetation types (Table 4.2). Only after the Sonoran and

98 Historic Vegetation Change

Table 4.2. Z-Scores for the Total Sample and for the Total Sample Minus Chihuahuan and Sonoran Desertscrub

Sample	Z Score log(x)
Total sample	−3.35*
Grassland	−1.18
Sonoran desertscrub	−1.947
Evergreen woodland	−1.442
Chihuahuan desertscrub	−2.272*
Total sample less Chihuahuan desertscrub	−2.61*
Total sample less Chihuahuan and Sonoran desertscrub	−1.867
Chihuahuan and Sonoran desertscrub	−3.053*

*Statistically significant at the .05 level.

Chihuahuan desertscrub data sets were removed from the total sample was the Z-score for the rest of the sample no longer significant. When the Sonoran and Chihuahuan desertscrub were considered by themselves, the Z-score was significant (see Appendix C).

The significant difference in the Sonoran and Chihuahuan desertscrub samples indicates an overall decrease in trees and large woody shrubs in all but eight of the twenty-three transects in Sonoran desertscrub and an overall decrease in trees and woody shrubs in all but two of the transects in Chihuahuan desertscrub. The decline in the numbers is due to human disturbance, particularly agricultural clearing, urban sprawl, rural subdivision development, and the expansion of other types of rural settlement. Decreases along eight of the transects in the Sonoran desertscrub sample are too small to be significant and may be the result of sampling error or are due to woodcutting and expanding rural settlement. Decreases along seven of the remaining transects are large, averaging 44 percent since 1935. All of the major decreases are due to subdivision development or the expansion of rural housing and agricultural clearing. Of the eight increases, six are too small to be significant, and two are large, averaging 47 percent since 1935–1937. The two major increases are due to invasions of mesquite in abandoned agricultural fields. In the Chihuahuan desertscrub sample, decreases along seven of the transects are too small to be significant and probably result from woodcutting; decreases along six of the transects are large, averaging 29 percent since 1935–1937. Four of the latter decreases are due to recent agricultural clearing and two are due to "mesquite conversion" programs. Increases along two transects are insignificant.

The transects indicate that there have been more decreases than increases in trees and large woody shrubs since 1935–1937, with sixty-nine transects showing decreases, thirty-eight showing increases, and four showing no change (Appendix B; see also Figures 4.15 and 4.16). Both the decreases and the increases appear to have been caused by human disturbance and land-management policies. For example, decreases are due largely to clearing, brush conversion, and woodcutting; increases, especially in the case of mesquite, are probably the result of grazing and fire suppression or, in the case of oak, the following of Forest Service controls on grazing, wildfire, and woodcutting. It should also be pointed out that the mesquite invasion of southeastern Arizona's rangelands may have peaked by 1940 and since that time, while mesquite has continued to increase in some areas, it has decreased in others because of expanding settlement, renewed demand for mesquite as fuel, and rangeland restoration programs.

That the most significant changes have been in Sonoran and Chihuahuan desertscrub is mainly because they are the vegetation types in which most of southeastern Arizona's settlement, especially irrigated agriculture, is located, and in terms of the percentage of total area cleared for settlement, they surpass the other major vegetation types.

My examination of changes in vegetation patterns and distributions in the 2,025 square miles covered by the matched aerial photography by overlapping the NHAP images onto the SCS images shows no shifts in the distribution of major vegetation types since 1935–1937 (Figures 4.13 and 4.14). Furthermore, there are no general changes in vegetation patterns between 1935–1937 and 1983–1984, except in cleared and overgrazed places or in certain areas where mesquite had increased greatly. The following directional vegetation changes can be identified in the repeat aerial photography spanning the period between 1935 and 1984: (1) an increase, particularly in some rangelands, of brush and scrubby trees, largely mesquite and acacia; (2) a steady increase in the area cleared for settlement, although there has been a decrease in farmland as some abandoned croplands have been reclaimed by exotics and native "pioneers"; (3) an expansion of exotic plants such as Lehmann's lovegrass and tamarisk; and (4) the continual extirpation of native riparian vegetation.

In the future, the annual accumulation of high-resolution vertical aerial photographs of southeastern Arizona will offer students of landscape change an inexpensive, reliable way to monitor changes in regional vegetation distribution.

CLIMATE VERSUS LAND USE

Climatic fluctuations are sometimes invoked to explain recent vegetation changes in North America (Bray 1971; Kline and Cottam 1979; Agee and Dunwiddie 1984; Tainter et al. 1984). Although climatic fluctuations clearly account for short-term vegetation changes in southeastern Arizona since the 1870s (Martin and Turner 1977; Bowers 1981; Gehlbach 1981; Glendening and Brown 1982; Goldberg and Turner 1986), climatic change, not climatic fluctuations, has been championed by Hastings and Turner (1965) and Neilson (1986) to explain the recent takeover of the rangelands by woody plants and the supposed upward retreat of plant ranges. Neilson concludes "that the 'pristine' vegetation of the northern Chihuahuan Desert, recorded 100 years ago, was a vegetation established under and adapted to 300 years of 'little ice age' and is only marginally supported under the present climate" (1986:33).

According to Sellers (1960) and Hastings and Turner (1965), there was a slight decrease in rainfall together with a slight increase in temperature in southern Arizona between 1898 and 1960 (see Figure 4.17).

An interesting twist to the theme of increasing aridity in southeastern Arizona has been put forth by Robert Balling (1988, 1989), a climatologist who suggests that the sharp vegetation discontinuity across the international boundary dividing Arizona and Sonora (Figure 3.1), due to contrasting land-use practices, has resulted in significant differences in summer maximum temperatures between the two states, with the Sonoran stations being significantly higher (by nearly 4.5°F.) than the Arizona stations. His finding indicates warming in desert areas where vegetation cover is decreasing and albedo is increasing. The differences in temperatures, however, appear to be associated more with differential evapotranspiration than with albedo changes.

Table 4.3 shows the variability of annual precipitation for five stations in southeastern Arizona from as early as 1866 to 1966; Table 4.4 shows drought and wet periods at Fort Lowell (near Tucson) from 1866 to 1961. Cooke and Reeves, who carefully reviewed all of the climatic data for southeastern Arizona and the analyses of Leopold (1951b), McDonald (1956), Sellers (1960), and Hastings and Turner (1965) conclude:

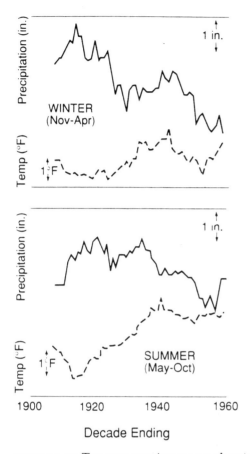

FIGURE 4.17. Ten-year running means showing mean seasonal precipitation at eighteen southern Arizona and western New Mexico stations (after Sellers 1960) with mean seasonal temperatures for the same eighteen stations (after Hastings and Turner 1965:282).

(1) there have been no statistically significant secular changes in annual, annual summer, or annual non-summer precipitation totals during the last hundred years at the stations studied in southern Arizona; (2) there has been a considerable variation in precipitation from year to year and ... drought and wet periods are a feature of the precipitation pattern; and (3) there has been an increase in the frequency of light rains and a reduction in the frequency of high-intensity rains. (1976:78)

Furthermore, they surmise that even if there has been a recent trend toward greater aridity, the changes in precipitation and temperature are so slight and protracted that it is unlikely they have modified the

Table 4.3. Variability of Annual Precipitation

	Mean Annual Precipitation (in.)	Standard Deviation	Coefficient of Variation	Years of Record
Fort Lowell-Tucson	10.991	3.291	0.299	100
Tombstone	13.378	4.445	0.332	70
Fort Grant	12.724	3.847	0.302	45
Fort Bowie	13.639	3.356	0.246	22
San Simon	8.886	3.162	0.355	60

SOURCE: Cooke and Reeves 1976:76.

Table 4.4. Drought and Wet Periods: Cumulative Precipitation Deficiency or Excess Between June and September at Fort Lowell, near Tucson

Drought	Cumulative Deficiency (inches below mean)	Wet Periods	Cumulative Excess (inches above mean)
1884–86*†	8.834	1866–69	5.72
1891–92	3.016	1871–72	11.47
1894–95*	3.606	1874–76	8.31
1899–1906*	10.794	1878–80	4.26
1912–13	3.866	1889–90	8.62
1915–16	0.966	1896–98	3.26
1932–34*	2.41	1907–11	7.83
1937–39*	3.45	1935–36	2.52
1944–45	1.90	1940–41	2.29
1947–49	8.44	1954–55	4.92
1951–53*	5.08		
1956–57	3.22		
1960–61	2.18		

SOURCE: Cooke and Reeves 1976:77.
*Drought terminated by two or more relatively wet summers.
†Preceded by incomplete data.

vegetation. Of course, there is also the possibility that anthropogenic impacts have resulted in these climatic trends! The regional variability of precipitation from year to year is well illustrated in Figure 4.18, which shows the annual deviation from mean precipitation for nineteen Arizona stations between 1898 and 1986, and Figure 4.19, which shows the same information for four stations in the study area.

No directional vegetation change since 1870 has been clearly linked to any trends, changes, or fluctuations in the climate. Potentially the most persuasive argument for climatic change is the claim by Hastings and Turner (1965) that there has been a recent upward displacement of vegetation ranges in southeastern Arizona and northwestern Sonora. The evidence, however, in the General Land Office surveyors' field notes, repeat ground and aerial photography, permanent plot studies, and historical descriptions of the nineteenth-century landscape does not support this claim. There is, of course, the possibility that directional changes in the ranges of the major vegetation types are too subtle

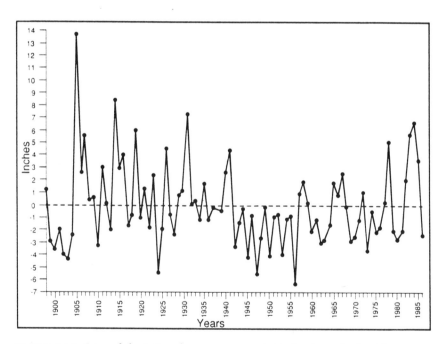

FIGURE 4.18. Annual deviation from mean precipitation (13.2 inches) for nineteen Arizona stations: 1898–1986. (The stations are Flagstaff, Holbrook, Jerome, Walnut Grove, Payson, Phoenix, Miami, Clifton, Gila Bend, Tucson, Bisbee, Parker, Willcox, Bowie, Grand Canyon, Kingman, Prescott, Williams, and Yuma.) Source: W. D. Sellers.

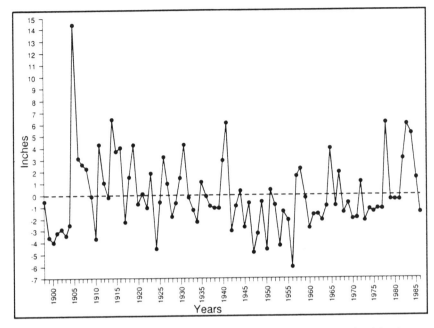

FIGURE 4.19. Annual deviation from mean precipitation (12.7 inches) for four stations (Tucson, Bowie, Willcox, and Bisbee) in the study area: 1898–1986. *Source*: W. D. Sellers.

to be detected in the aforementioned evidence or are too insignificant to be identified in such a short time span.

That the directional changes in the vegetation of southeastern Arizona since 1870 bear little or no relationship to increasing aridity is further supported by the investigations of Humphrey (1987). As noted earlier, Humphrey (1987:430) observes that almost all of the major vegetation changes along the U.S.-Mexico boundary between the Rio Grande and the Colorado River since the 1890s have occurred in the more mesic eastern section of the boundary, which is more densely populated and grazed, while the more arid, largely ungrazed, western part of the boundary has undergone little or no change. If change had occurred, it would be expected across several zones.

Furthermore, most of the directional changes were initiated at different times and in different places; the only consistent pattern is that they always follow some type of human disturbance. For example, most of the mesquite increase has taken place since the 1880s in areas that have been heavily grazed, have been subjected to few or no fires, or have been cleared and abandoned. It should be pointed out, however, that mesquite is continuing to increase in some protected areas, possibly due to autogenic succession (Archer et al. 1988) and fire exclusion.

Probably more time has been spent on massaging the climatic change hypothesis than on any other factor of vegetation change, and yet it remains the least convincing. Perhaps because we are taught that climate is the primary factor controlling the distribution of the earth's biomes, we continue to search the climatic record for explanations of recent vegetation changes, especially those of a regional nature.

It is an old axiom in geography that the modern landscape reflects past and present cultural perceptions and uses of the land as influenced by the physical environment. Only by understanding human-land relations historically and ecologically can we come to truly understand the evolution of the present wild landscape and differentiate between the natural and the anthropogenic factors of change. To obtain a better perspective on the historical-ecological relations between humans and the vegetation of southeastern Arizona since 1870, the following chapters examine the major historic land uses that have effected the evolution of the present wild landscape.

PART II
Primary Historic Human Impacts Since 1870

Many of the major directional changes in the vegetation of southeastern Arizona were set in motion around 1900 and, according to the record, continued a degradation pattern established during the 1880s and 1890s. This part of the book is devoted to identifying and analyzing the major historic land uses that either point to or have resulted in major vegetation changes after 1870. Emphasis is placed on livestock grazing, wildfire exclusion, fuelwood cutting, exotic plant introductions, agriculture and groundwater pumping, logging, and wild hay harvesting. The emphasis is on those land uses which initiated and perpetuated vegetation changes in the 1870s and not on contemporary land-use impacts.

5 Livestock Grazing

Few areas in southeastern Arizona, no matter how isolated, have escaped the impact of cattle ranching, but we know surprisingly little about the role of livestock in the evolution of the wild landscape or in most of the identified vegetation changes since 1870. Generally, only those vegetation changes (largely in the grasslands) which have resulted in lower range-carrying capacities have been identified; even then, the relation of cattle to some of those changes is uncertain. Were cattle, for example, responsible for woody plant increases in the grasslands? Many landscape changes in southeastern Arizona, as well as in the rest of the American Southwest, closely followed the arrival of large-scale Anglo-American ranching and the overgrazing prevalent near the turn of the century (see Figure 5.1).

In 1901, D. A. Griffiths, chief botanist in charge of grass and forage plant investigations for the Arizona Experiment Station in Tucson, observed that the rangelands of southern Arizona were more degraded than any others he had seen in the western United States (1901:9). This observation was made at a time when livestock numbers had been greatly reduced as a result of the 1891–1893 drought (Cameron 1896:224). To obtain accurate knowledge of range conditions before the livestock boom of the 1880s, Griffiths (1901:11–14) devised a circular and sent it to a select group of pioneer ranchers. The answers to the questions on the circular by H. C. Hooker, proprietor of the Sierra Bonita Ranch, and C. H. Bayless, owner of a large ranch near Oracle, better explain the impact of cattle on vegetation changes in southeastern Arizona in the 1880s and 1890s than much of the research published since.

Circular questions (Q) and answers by H. C. Hooker (H) and C. H. Bayless (B) follow:

1. (Q) With what portions of the Territory are you especially familiar?
   (H) The southeastern.
   (B) The San Pedro Valley and southern part of Pinal County.
2. (Q) How long have you been acquainted with the regions spoken of in question 1?
   (H) Thirty-five years.
   (B) Fifteen years.
3. (Q) What was the relative abundance of the feed on the native range at the time you first became acquainted with it, compared with the present time?
   (H) Fully double.
   (B) At that time ten animals were kept in good condition where one now barely exists. However, those ten animals were then rapidly destroying the vegetation, not making proper use of it.
4. (Q) Will you please compare the grazing conditions in two or more regions with which you are familiar: for instance, the Santa Cruz, San Pedro and Sulphur Spring valleys?
   (H) These regions have been diminished in grazing facilities fully 50 percent in twenty-five years.

FIGURE 5.1. Cattle roundup in the Sulphur Springs Valley circa 1890. Photograph by C. S. Fly, courtesy of the Bisbee Mining and Historical Museum, Brophy Collection.

5. (Q) Can you describe any specific instances of the destructive action of water in gullying out the river valleys? Can you state how and at what time such gullying started in any particular instance, and the extent to which the washing progressed in a given time?

(H) The San Pedro Valley in 1870 had an abundance of willow, cottonwood, sycamore, and mesquite timber; also large beds of sacaton and grama grasses, sagebrush, and underbrush of many kinds. The river bed was shallow and grassy and its banks were beautiful with a luxuriant growth of vegetation. Now the river is deep and its banks are washed out, the trees and underbrush are gone, the sacaton has been cut out by the plow and grub hoe, the mesa has been grazed by thousands of horses and cattle, and the valley has been farmed. Cattle and horses going to and from feed and water have made many trails or paths to the mountains. Browse on the hillsides has been eaten off. Fire has destroyed much of the shrubbery as well as the grass, giving the winds and rains full sweep to carry away the earth loosened by the feet of the animals. In this way many waterways have been cut from the hills to the river bed. There is now little or nothing to stop the great currents of water reaching the river bed with such force as to cut large channels and destroy much of the land under cultivation, leaving the river from 10 to 40 feet below its former banks. Thus it has caused much expense in bringing the water to the cultivated lands, and necessitated much labor to dam up the channel and keep the irrigating ditches in repair.

(B) About twelve years ago the San Pedro Valley consisted of a narrow strip of subirrigated and very fertile lands. Beaver dams checked the flow of water and prevented the cutting of a channel. Trappers exterminated the beavers, and less grass on the hillsides permitted greater erosion, so that within four or five years a channel varying in depth from 3 to 20 feet was cut almost the whole length of the river. Every year freshets are carrying away new portions of the bottom lands. At present this valley is a sandy waste from bluff to bluff, while the few fields remaining are protected from the river at large and continuous expense. Thus, in addition to curtailing the area of good land, the deep channel has drained the bottoms, leaving the native grass no chance to recover from the effects of close pasturing. It also makes it more difficult to get irrigating water onto the surface of the land.

6. (Q) What influence has this gullying had on the productiveness of the river bottoms?

7. (Q) What grasses or other native forage plants furnish the greatest amount of feed at the present time in your vicinity? (If you do not know the names of these plants and are willing to send us samples, so state in answer to this question, and we will send you franks so that you can forward the samples to us free of charge.)

(H) Gramas, sacatons, bunch, and six-weeks grasses.

(B) Of the rich grama grasses that originally covered the country so little now remains that no account can be taken of them. In some parts of the foothills alfilaria furnishes limited but excellent pasture during the spring and early summer. Where stock water is far removed some remnants of perennial grasses can be found. Grasses that grow only from seed sprouted by summer rains are of small and transitory value. The foliage of the mesquite and catsclaw bushes is eaten by most animals, and even the various cacti are attempted by starving cattle. However, the thorns and spines of the cacti more than offset the value of the pulp. No better pasture was ever found in any country than that furnished by our native grama grasses, now almost extinct.

8. (Q) Do you attribute the present unproductive condition of the range to overstocking, drought, or to both combined? Please explain why.

(H) Principally to overstocking. In times of drought even the roots are eaten and destroyed by cattle, while if not fed down or eaten out the roots would grow again with winter moisture.

(B) The present unproductive conditions are due entirely to overstocking. The laws of nature have not changed. Under similar conditions vegetation would flourish on our ranges today as it did fifteen years ago. We are still receiving our average amount of rainfall and sunshine necessary to plant growth. Droughts are not more frequent now than in the past, but mother earth has been stripped of all grass covering. The very roots have been trampled out by the hungry herds constantly wandering to and fro in search of enough food. The bare surface of the ground affords no resistance to the rain that falls upon it and the precious water rushes away in destructive volumes, bearing with it all the lighter and richer particles of the soil. That the sand and rocks left behind are able to support even the scantiest growth of plant life is a remarkable tribute to our marvelous climate. Vegetation does not thrive as it once did, not because of drought, but because the seed is gone, the roots are gone, and the soil is gone. This is all the direct result of overstocking and can not be prevented on our open range where the land is not subject to private control.

9. (Q) Will you please state the largest number of cattle which, in your opinion, have at any time grazed on any particular range with which you are acquainted, and at what time? What do you estimate is the present carrying capacity of the same range?

(H) There were fully 50,000 head of stock at the head of Sulphur Spring Valley and the valley of the Aravaipa in 1890. In 1900 there were not more than one-half that number and they were doing poorly.

(B) Twelve years ago 40,000 cattle grew fat along a certain portion of the San Pedro Valley where now 3,000 can not find sufficient forage for proper growth and development. If instead of 40,000 head 10,000 had been kept on this range, it would in all probability be furnishing good pasture for the same number today. Very few of these cattle were sold or removed from the range. They were simply left there until the pasture was destroyed and the stock then perished by starvation.

10. (Q) Provided we should be able to furnish seed, would you be willing to put it in the ground in proper shape in some favorable situation on your place where cattle will not graze it for at least one year after planting? A very small patch would be required, say 50 feet square. Such an experiment would enable us to determine what forage plants are best adapted to your locality.

(H) I will place 1 acre or more under fence on my land in any situation you may select for your experiments, providing you will superintend the planting and direct the cultivation, taking from my ranch such teams, farming tools, employees, etc., as you may require. I am, respectfully yours,

H.C. Hooker,
Proprietor Sierra Bonita Ranch

(B) Yes, I will do so gladly. Object lessons of this kind will prove conclusively that overstocking, not drought, has made our country a desert.

C.H. Bayless

The degraded range conditions near the turn of the century are further substantiated in pictures of southeastern Arizona in *Views of the Monuments and Characteristic Scenes Along the Boundary Between the United States and Mexico West of the Rio Grande 1892–1895* (International Boundary Commission 1898)[1] and in the Roskruge Photograph Collection at the Arizona Historical Society in Tucson. In those pictures, hundreds of square miles of rangeland are denuded of cover; the grasses, even the sacaton in the bottomlands, are grazed to the ground; the hills are covered with cattle trails (terrasettes); erosion is rampant; and the oaks and other trees have browse lines. In 1891, J. W. Toumey, the botanist who preceded Griffiths at the Arizona Experiment Station, wrote: "There are valleys [in southeastern Arizona] over which one can ride for several miles without finding mature grasses sufficient for herbarium specimens without searching under bushes or in other similar places" (1891b).

It is well known that many areas of grassland in many parts of the world have been converted to chamaephyte and scrub vegetation because of heavy grazing for several centuries. According to Zohary (1973), much of the Fertile Crescent was once grassland, but centuries of heavy grazing have extirpated the perennial grasses. Similarly, heavy grazing has eliminated perennial grasses from the South African veld (Acocks 1953; Werger 1980), southeastern Rhodesia (Kelly and Walker 1976), eastern Botswana (Van Vegten 1983), the Rio Grande Plain of

Texas (Lehman 1969), California's Central Valley (Burcham 1957), and Chile's Valle Central (Bahre 1979).

HISTORY OF THE LIVESTOCK INDUSTRY

The history of the livestock industry in southeastern Arizona is well chronicled by Cameron (1896), Haskett (1935), Morrisey (1950), and Wagoner (1951, 1952, 1961). Large-scale cattle ranching has been carried on in the area since the 1870s, although cattle and other livestock were introduced into southeastern Arizona two centuries earlier (Bolton 1936:269). While the history of ranching in Arizona is muddled during the Spanish and Mexican occupations, there is evidence that large numbers of cattle, horses, sheep, goats, burros, and mules may have been in the region from 1700 to 1840, especially in the 1820s and 1830s, when large Mexican land grants were established within sixty miles of the present international boundary (Cameron 1896; Mattison 1946). These grants were the San Rafael de la Zanja, María Santísima del Carmén, Luis María Baca, San Bernardino, San Ignacio de la Canoa, San José de Sonoita, San Ignacio del Babocomari, San Juan de las Boquillas y Nogales, and San Rafael del Valle (see Figure 2.10). Supposedly thousands of horses, cattle, mules, and sheep were run on these grants (140,000 cattle on the Babocomari and San Bernardino grants alone) (Bartlett 1854:vol. 1, 396; Haskett 1935:6). Mariana Diaz, a native of Tucson who in 1873 was more than 100 years old, said that "the country around Tucson was covered with horses and cattle in the past and that the trails were so plentiful that it was quite inconvenient to get through the immense herds . . . and that they [cattle] were valuable only for hides and tallow" (*Arizona Weekly Citizen* July 21, 1873). Nevertheless, Apache depredations from 1692 to 1786 and from the late 1820s to 1872 greatly hindered ranching (Cameron 1896; Nentvig 1980).

Considering the general lack of livestock water developments during the early nineteenth century in Arizona (there were no windmills or stock tanks) and the intermittent nature of most streams, it is difficult to believe that the grass and browse in the rangelands adjacent to major sources of perennial water could have supported such high numbers of cattle. Furthermore, large-scale cattle ranching during the 1820s would have been curtailed by Apache depredations that presumably brought ranching operations to a standstill during this period (Haskett 1935). Even if there had been large numbers of cattle in the region in the 1820s and 1830s, there is no evidence of overgrazing. Had the ranges been overgrazed, one might seriously question the premise that overgrazing led to the stream entrenching, fire exclusion, and brush invasion that occurred after 1890. That overgrazing was insignificant in the 1820s and 1830s is also substantiated by the fact that most descriptions of southeastern Arizona from 1850 to 1880 emphasize largely pristine vegetation ideal for cattle (Hastings and Turner 1965:35–50).

Between 1846 (when Lieutenant Colonel Philip St. George Cooke led the Mormon Battalion through southeastern Arizona during the Mexican War) and the Gadsden Purchase in 1853, there were a number of accounts of wild cattle in the region, especially in the San Bernardino and San Pedro valleys (Hastings 1959; Hastings and Turner 1965:34). For example, Cooke, whose battalion was attacked by wild cattle at the junction of Babocomari Creek and the San Pedro River, noted, "There is not on the open prairies of Clay County, Missouri, so many traces of the passage of cattle and horses as we see every day" (1938:79, 143). When John Russell Bartlett, commissioner to the United States-Mexican Boundary Survey, entered southern Arizona with the U.S.-Mexican Boundary Survey in 1851, he described the San Bernardino Valley as desolate and covered with cattle trails. Furthermore, near present-day Douglas, his party used cattle dung for cooking fires because of the lack of firewood in the area. Bartlett also noted a party of thirty to forty Mexicans camped at the confluence of Babocomari Creek and the San Pedro River hunting wild cattle (Bartlett 1854:vol. 1, 398). Apparently, long after the grants were abandoned, Mexicans continued to come to the area to hunt cattle for tallow, hides, hooves, and meat (Wagoner 1952:128). In addition, from 1850 to 1853 large numbers of cattle were driven across southeastern Arizona by immigrants to California but few, if any, of these cattle appear to have been left behind (Cameron 1896; Brady n.d.:39). They must, however, have affected the vegetation along the major trails. Christiansen (1988:95) notes that there may have been 100,000 wild cattle in southeastern Arizona during the 1840s and 1850s, but there is only circumstantial evidence for this figure.

Although large numbers of cattle were driven into southeastern Arizona to meet government and local needs after 1866 (Haskett 1935:19), large-scale cattle ranching did not develop until nearly a decade after the Civil War. In 1872, H. C. Hooker, the most prominent Anglo rancher in southeastern Arizona, had 11,000 cattle in the Sulphur Springs Valley (Haskett 1935:23; Morrisey 1950:152). At that time, small numbers of cattle were also in the Santa Cruz Valley south of Tucson, in parts of the San Pedro Valley, along Babocomari Creek between its junction with the San Pedro River and present-day Sonoita, along the Gila River from Duncan to Thatcher, and along Sonoita Creek west of present-day Patagonia to Calabasas (Haskett 1935; Morrisey 1950; Wagoner 1961).

The boom in cattle ranching began when the Southern Pacific Railroad came to southeastern Arizona. In addition, the spread of the windmill and the removal of the Apache threat opened up the region to ranching. In 1881 the Southern Pacific advertised for settlers; soon after, ranchers from overgrazed areas in Texas, New Mexico, and the Mexican states of Durango, Chihuahua, and Sonora began moving their herds into southern Arizona. The transformation of the rangelands occurred so rapidly that by 1884, according to Cameron (1896), a pioneer rancher in the San Rafael Valley, every running stream and permanent spring had been claimed and adjacent ranges stocked with cattle. In the U.S. Congress, *Annual Report of the Governor of Arizona* (1883:508), Governor F. A. Tritle boasted that there were 34 million acres of grassland in the Arizona Territory—enough pasture, he claimed, for 7,680,000 cattle. Then, as now, the rangelands with the highest carrying capacity were found in the southeastern part of the state (Hecht and Reeves 1981:51).

Between 1885 and 1890, large investments continued to pour into southeastern Arizona's cattle herds, and by 1890 the entire region must have looked like one huge cattle ranch. The *Arizona Daily Star* (Jan. 5, 1890) reported: "Conservative cattlemen estimate that at the lowest calculations there are now more than 150,000 head of cattle in Pima County.... others claim as many as 250,000." The *Arizona Daily Star* (June 19, 1888) listed the Pima County cattle barons and their livestock as follows:

> Pima County cattle barons ... W. L. Vail and Associates have not less than 20,000 head. Last year they branded over 4,000 calves. Maish and Driscoll have in the neighborhood of 23,000 head and they also branded over 4,000 calves last year. The Land and Hays Company of the Barbacumbrie [Babocomari] grant, have about 15,000 head, a number of which are in Cochise County. They branded 3,000 calves. The Cameron Bros. on the San Rafael Ranch, have almost 12,000. They handled nearly 3,000 calves last season. Sabino Ortero has about 7,000 head and handled over 2,200 calves. Gen. Pusch has about 6,000 head. Richardson and Gormley have between 7,000 and 8,000 head. They branded 2,100 calves. They recently purchased 100 head from Ashburn Bros. Arnatha has about 2,500 head on his Santa Cruz ranch, and there are about a dozen others in the Santa Cruz Valley who have from 500 to 1,500 head. Then, there are a number on the San Pedro who have from 500 to 1,000 head, and also a number of stockmen through the Tanque Verde, Santa Ritas, Sierias [Sierritas] and ranges west of Tucson, who have bunches of cattle ranging from 500 to 1,000 head.

Before 1892, sheep were also important in southeastern Arizona and, according to census reports, actually outnumbered cattle. The *Arizona Weekly Star* (Jan. 1, 1879) reported 78,500 sheep in Pima County, compared with 68,600 cattle. In 1880, the Ciénega Ranch just west of the Whetstone Mountains was reported to have 23,000 sheep (U.S. Bureau of the Census 1883a), while the *Arizona Daily Star* (Mar. 20, 1892) noted 20,000 sheep in the Chiricahua Mountains. Between the late 1870s and early 1890s large numbers of sheep were also in the Sonoita Valley, Santa Rita Mountains, lower Santa Cruz Valley, and along the west side of the Chiricahua Mountains (*Arizona Weekly Citizen* Dec. 6, 1873, June 6, 1874, Dec. 5, 1874, Mar. 13, 1875, Aug. 4, 1877; *Arizona Daily Star* Oct. 22, 1884, Mar. 29, 1892, May 28, 1892; Potter 1902). Haskett (1936) also noted large numbers of sheep on Mexican land grants in the 1820s and 1830s. What impact these sheep had on southeastern Arizona rangelands, however, is unrecorded.

In U.S. Congress, *Annual Report of the Governor of Arizona* (1893), Governor J. N. Irwin listed 121,377 cattle in 1891 for Pima County, which then included present-day Santa Cruz County, and 95,850 cattle for Cochise County. These figures, conservative because few ranchers reported the total numbers of their cattle to the tax assessors, appear at the beginning of the 1891–1893 drought (Haskett 1935). The year 1890 was deficient in rainfall, and 1891 and 1892 were almost devoid of summer rains. In the first months of 1893, the combined effect of drought and overgrazing led to the death of an estimated 50 to 75 percent of the livestock, mostly in southeastern Arizona. Summer rains in July 1893 saved the cattle industry from complete ruin, but overstocking and overgrazing continued. Hendrickson and Minckley (1984:161) estimate that, shortly before and during the 1891–1893 drought, 377,474 cattle grazed southeastern Arizona, more than twice the number of cattle presently grazing the rangelands of Pima, Cochise, Santa Cruz, and Graham counties. According to Gray (1940:131–132):

> ... the years following the drought brought forth a new vocation, which the cowboys looked upon with much disgust—almost akin to grave robbing. That was the business of the sun-bleached bones of the drought victims. Near almost every railroad station there were accumulated great stacks of bones hauled in from cattle ranges. Men appeared with wagons and started to gather up bones and horns of dead cows, and soon it was a common sight to see piles of white bones growing daily in size at every shipping point. ...

After 1893 major changes in the landscape occurred: many areas were completely denuded of grass cover and the topsoil was eroded; *ciénagas* were destroyed and sections of the San Pedro, San Simon, Santa Cruz, and Babocomari rivers ceased being perennial; headward cutting and channel entrenching began along most of the major streams; the hills were covered with cattle trails; oaks and other trees had browse lines; and woody xerophytic shrubs and weeds began to increase and/or invade the rangelands (U.S. Congress, House 1893:319).

Between 1893 and enactment of the Taylor Grazing Act in 1934, the entire region—with the possible exception of some forest reserves—was continuously overstocked and overgrazed, but never again were livestock numbers anywhere near the levels of 1891 (Wagoner 1952). Until 1900, cattle moved at will because most of the area was open range. The only major fences followed railroad rights-of-way. Even most National Forest lands were not fenced until the 1930s, and public lands were among the most abused. The introduction of fences, however, was resisted by some ranchers, and even today some old-timers say that fencing the open range resulted in lower carrying capacities because cattle were concentrated in limited areas and worse overgrazing resulted. According to the old-timers, "The cattle went where the feed was when there was open range, whereas today with fences and supplemental feeding, the cattle stay in pastures far longer than the grass can feed them, thus ruining the land" (Bahre 1977:27).

The Stock Raising Act of 1916, which allowed cattle homesteads of 640 acres (Rodgers 1965:96), closed the open range era in southeastern Arizona because, with increased homestead activity, cattlemen were gradually forced to change their methods of operation. Previously, most ranchers depended upon small landholdings as a base of operations and grazed their cattle on the open ranges. An increase in the number of ranching homesteads, however, led to a decline in the amount of open range and forced the large ranchers to accumulate more patent land to ensure adequate grazing for their cattle. After 1920, the number of ranches with more than 1,000 acres increased greatly (Rodgers 1965: 109).

Overgrazing of Forest Service lands was largely curtailed after 1906, although overgrazing continued on Bureau of Land Management (BLM) lands until the enactment of the Taylor Grazing Act in 1934. The intent of this act was to prevent overgrazing, and consequent soil and watershed deterioration, as well as to stabilize the livestock industry dependent upon public range (Voigt 1976). Overstocking, however, has continued until today on some private and public lands.

The Forest Service and other governmental agencies initiated a series of federally sponsored livestock programs that have led to a new set of changes in the wild lands, generally at the expense of those plants least palatable to *Bos*. For example, the Forest Service and BLM initiated contour plowing, fencing, and rotational and deferred grazing, and prescribed burning, fire suppression, control of woody plants by toxic chemicals, introduction of exotic forage plants, construction of check and spreader dams, predator and rodent control, weed control, and chaining or bulldozing of oaks, mesquite, juniper, and pinyon. These practices and a host of other activities designed to protect watershed and improve the livestock industry have culminated in much of the landscape change witnessed today and have resulted in at least short-term alterations in the vegetation cover over large parts of southeastern Arizona.

EFFECTS OF CATTLE AND LIVESTOCK MANAGEMENT

One major change in the grasslands of southeastern Arizona (as well as in other grasslands of the American Southwest) is the dramatic increase in scrubby, xerophytic trees and shrubs since the late 1890s (Thornber 1910; Leopold 1924; Parker and Martin 1952; Humphrey 1956, 1958; Humphrey and Mehrhoff 1958; Harris 1966; Wright 1980). These woody increasers, the spread of which has been influenced, if not caused, by overgrazing, and which are relatively unpalatable to livestock (Humphrey 1953; Schmutz et al. 1968), predominantly include mesquite, acacia, creosote bush, juniper, oak, burroweed, senecio, and snakeweed. Darrow (1944) estimates that by 1944 mesquite and acacia had invaded 1.7 million acres of grassland in Cochise County alone, and Parker and Martin (1952) report that by 1952 fully half of the 9 million acres of the land supporting mesquite in southern Arizona had been colonized by that species since 1850. In addition, burroweed and snakeweed had increased dramatically on overgrazed ranges by 1937; each at that time dominated nearly 5 million acres of former grassland (Upson et al. 1937; Humphrey 1937, 1949; Tschirley and Martin 1961). Less widely recognized is the role of overgrazing in the increase of exotic annual plants in southeastern Arizona. According to Thornber (1907, 1910) and Griffiths (1910), rangelands were so degraded by 1900 that both native and exotic annual grasses and forbs were replacing native perennial grasses.

Livestock promote shrub invasion and increase in at least four ways:

1. They disseminate viable seeds in their droppings, hair, and hooves (Glendening and Paulsen 1955) and scarify the seeds of plants, such as mesquite, in their alimentary tracts (Reynolds and Glendening 1949; Martin and Cable 1974:3). Nevertheless, mesquite continues

to increase in certain areas where livestock grazing has been eliminated, even in good stands of perennial grasses (Glendening 1952; Branscomb 1956, 1958; Cable 1967; Cable and Martin 1973; Archer et al. 1988).
2. Livestock promote weeds and shrubs through opening up the grasslands by (a) weakening grass vigor and cover, thus reducing the ability of grass to compete with woody shrubs, and (b) by exposing the topsoil to erosion, which results not only in its removal but also in grass-barren swales and pedestaling (Haskell 1945; Brown 1950; Glendening and Paulsen 1955; Cable 1969; Martin 1975; Wright et al. 1976; Wright 1980:6; Walker et al. 1981; Hennessy et al. 1983). Experiments at the Santa Rita Experimental Range demonstrated that sixteen times as many mesquite seedlings were established on bare areas as in vigorous stands of perennial grasses (Glendening and Paulsen 1955). Moreover, once established, mesquite seedlings are severely restricted in good stands of grass.
3. Grazing compacts the topsoil, which not only causes soil creep (terrasettes) on sloped land but also reduces the moisture content in the upper layers of the soil. The latter favors trees such as mesquite, which can draw on deeper water supplies.
4. Grazing reduces grass cover and thereby decreases the fine fuel load and the incidence of wildfire (Leopold 1924; Humphrey 1958, 1962; Martin 1983:604).

Most ecologists and range managers believe that the takeover of the rangelands by woody plants is the result primarily of overgrazing in conjunction with drought and fire suppression policies. According to them, the introduction of large numbers of cattle in the late 1870s and early 1880s led to severe overgrazing and the consequent removal of fuel, thus checking or preventing the widespread wildfires that once held the brush and trees in check (Toumey 1891b; Griffiths 1910; Wooton 1916; Leopold 1924; Humphrey 1958, 1962; Martin 1975; Wright and Bailey 1982). However, just grazing in its own right may have resulted in the increase because mesquite, burroweed, snakeweed, desert broom, and several acacias quickly increase on heavily grazed sites.

Oak and juniper have also reportedly invaded or increased in the upper-altitude margins of the grasslands bordering the evergreen woodlands. The Forest Service, the BLM, and most ranchers view these increases as a threat to forage production, and in some areas of southeastern Arizona, oaks and junipers have been chained, burned, and bulldozed

(Bahre 1977). Whereas cattle are thought to have a negative effect on oak regeneration, they are often identified as a major cause of increased coniferous forest density (Runnell 1951; Cooper 1960; Blackburn and Tueller 1970; Madany and West 1983) and as a reason for invasion by trees of high mountain meadows (Vale 1977, 1981). Juniper increases, although occurring in some parts of southeastern Arizona, are a greater problem in northern Arizona, where attempts to eradicate juniper from rangelands have been going on at least since 1940 (Johnsen and Dalen 1984).

Another major landscape change since 1870 in southeastern Arizona and the American Southwest, also often attributed to livestock overgrazing, is accelerated arroyo cutting. Arroyo cutting and stream channel deepening in valleys of southeastern Arizona occurred from 1850 to 1920, but especially from 1870 to 1890 (Cooke and Reeves 1976:94, 187). Supposedly overgrazing, along with trampling and soil compaction, increased runoff; this resulted in increased discharges, greater stream velocities, increased stream capacities to do work, and, eventually, channel deepening and headward cutting. Because heavy overgrazing of range coincides with other land uses, it is difficult to sort out the role of grazing in stream deepening.

In addition, cattle have affected riparian tree regeneration, riparian community stability, stream sediment loads, stream temperatures, aquatic life, and stream bank stability.[2] Forbes commented in 1902 on the negative impact of overgrazing on the Gila River watershed:

> The runoff of the Gila has been greatly affected within recent years by the operations of stockmen upon its watershed. Consisting of at least 90 per cent of open and originally grass-covered country, the region has at times been so heavily over-stocked that the grasses have everywhere been depleted, and in some districts practically destroyed. The rains, coming upon these bared and trampled ranges, fall with but little obstruction into the watercourses, giving rise to sudden flood and violent floods of great erosive power, which carry enormous quantities of sedimentary matter. (1902:183)

> The exceedingly muddy character observed [in the waters of the Gila River] is without doubt quite recent, according to the testimony of those who have irrigated many years under this stream, being caused by the excessive numbers of cattle at times permitted on the grassy watersheds of this river. Cause and effect are here so directly connected as to permit of no question, as when a cattle trail deepens into a gully with running water, or as when an overgrazed valley bottom is converted into bad-lands by the combined action of cattle and flood waters. Erosion of this character, destroying the grazing country, changing the nature of the water supply, and introducing an element of doubt into prospective storage reservoirs, is at once disastrous to stockmen, farmers, and investors. (1902:187)

In order to improve the rangelands for livestock or simply to revegetate degraded grassland, large areas of scrub-mesquite have been cleared and reseeded with exotic forage grasses.

Big sacaton (*Sporobolus wrightii*), once an important native forage plant favoring bottomlands and riparian areas, now occupies less than 5 percent of its original range (Humphrey 1958; Cox 1984b). Formerly important for hay making (Bahre 1987), it has been reduced mostly by disturbances other than grazing. In the past, ranchers frequently burned the old coarse growth because sacaton makes good forage when it is green. The Coronado National Forest now recommends that ranchers mow sacaton rather than burn it to lessen the risk of large conflagrations (Cox and Morton 1986).

To appreciate how the wildlands of southeastern Arizona have been affected by range management practices, one need only examine the conservation plans, permittee plans, and range inspection reports for cattle allotments in the Coronado National Forest. These plans contain advice on mowing sacaton, controlling rat and pocket gopher infestations, placing salt blocks, reseeding forage plants, developing water sources, building fences, and many other ways to "improve" the rangelands. This range management philosophy is exemplified by Herbel (1985:21) who writes:

> A positive approach is needed to replace unwanted species with those that are more useful to man. Useful practices may include: (1) mechanical, chemical and/or biological control of unwanted plants; (2) fires to control unwanted plants where fuel is adequate; (3) revegetation to replace unwanted plants with useful plants; and (4) introduction of different animal species to use the range ecosystems more efficiently.

These and other range improvement programs have introduced changes in the landscape, many of them unidentified, much less studied.

The BLM's environmental impact report in 1978 for the Upper Gila-San Simon area (the entire San Simon watershed and the lower San Pedro Valley from the 2nd Standard Parallel South north to the Gila River) exemplifies both the real and the perceived changes due to cattle and livestock management programs in southeastern Arizona. The Upper Gila-San Simon study area comprises 2,804,712 acres (4,382 sq. mi.), of which 552,960 acres are privately owned, 902,071 acres are in state trust, and 1,349,681 acres are under federal control. The San Simon Valley, often regarded as one of the most seriously disturbed environments in the United States (Peterson 1950:410), has been a focus of land management and conservation efforts since 1950. A stark testament to the long period of overstocking and overgrazing, the valley is characterized by extensive soil erosion and gullying, impoverished vegetation, and an annual sediment yield of as much as three acre-feet per square mile over 10 percent of the area (Jordan and Maynard 1970). Historical descriptions emphasize that in the 1880s the valley's vegetation and other conditions were ideal for cattle—meadows were covered with two-foot-high lush grasses, open areas were dominated by gramas, and abundant water was found on the valley floor (Hinton 1878; *Arizona Daily Star* Nov. 29, 1882; Barnes 1936). More than 50,000 cattle grazed in the valley in the 1880s (Thornber 1910), whereas at present fewer than 5,000 graze there.

Today, at least 10 percent of the Upper Gila-San Simon area suffers from critical or high erosion; 32 percent suffers from moderate erosion. Thus, almost 42 percent of the area (1,151,412 acres) suffers moderate to high erosion (U.S. Bureau of Land Management 1978). Since Anglo settlement, nearly all of this seriously eroded land has never been used for anything other than livestock grazing! Furthermore, 37 percent of the vegetation cover was rated in poor condition (less than 25 percent of the original cover is present) and 41 percent was rated fair (less than 50 percent of the original cover is present) (ibid.). Of the rangelands, 91 percent were judged to be in poor or fair condition (ibid.). Range condition reflects the present state of vegetation in relation to its assumed natural potential. It should be kept in mind, however, that the determination of the natural potential of plant communities is largely subjective. According to Martin (1975), the rangelands in southern Arizona have been managed for cattle for so long that we are uncertain about their pregrazing condition.

That different range management practices result in different impacts on the vegetation can be clearly seen in the spatial contrast in vegetation across the Arizona-Sonora boundary (Bahre and Bradbury 1978) (see Figure 3.1). Although other land-use differences account for some contrasts in the vegetation spanning the boundary, the major cause has been disparate perceptions and management of grazing in the United States and Mexico.

Probably no single land use has had a greater effect on the vegetation of southeastern Arizona or has led to more changes in the landscape than livestock grazing and range management programs. Undoubtedly, grazing since the 1870s has led to soil erosion, destruction of those plants most palatable to livestock, changes in regional fire ecology, the spread of both native and alien plants, and changes in the age structure of evergreen woodlands and riparian forests.

6 Fire

Before the 1890s, wildfires were common in southeastern Arizona wherever there was sufficient fuel. The Mogollon Rim country north of the Gila River has among the highest incidences of lightning fires in the United States (Komarek 1968; Schroeder and Buck 1970; Brown and Davis 1973; Pyne 1984), and 73 percent of all fires since 1959 in the Coronado National Forest have been started by lightning (Table 6.1, Figure 6.1). In addition, the Coronado National Forest leads the Southwest in average annual acreage burned by lightning fires (Barrows 1978). The lightning fire season in southern Arizona begins in May, peaks in July, and runs into October (ibid.). The zone of maximum lightning fire occurrence by elevation in southern Arizona appears to be above 6,000 feet (Figure 6.2) (Barrows 1978:65; Baisan 1988:12).

Lightning fires undoubtedly have occurred for millennia, and wildfires have been set by humans ever since they arrived in the region. The significance of wildfires in the vegetation history of southeastern Arizona is attested to by (1) the abundance of fire scars in the tree-ring record (Weaver 1968, 1974; Baisan 1988; Swetnam et al. 1989, Swetnam in press); (2) rapid postfire recovery in most of the major vegetation types (Humphrey 1958; Cable 1967; Weaver 1968, 1974; Bock et al. 1976; Pase and Granfelt 1977; Wright and Bailey 1982); (3) historic records of wildfires and burning by Amerinds (Dobyns 1981; Pyne 1982; Bahre 1985); and (4) the high incidence of lightning ignitions in summer.

When Anglo-American settlers arrived, however, the fire regime changed. Consequently the structure and, in some cases, the composition of the vegetation changed. Pyne (1984:253) describes the new fire regime thus:

> With heavy grazing, with the reservation of virtually all forest lands into protected areas during the late nineteenth century, with demands for watershed protection by irrigation agriculturalists, deliberate fire control came to the Southwest, and with it a new fire regime. As in the Northern

Rockies, this regime was not shaped by adding new ignition sources so much as by removing old ones. In some instances, active fire control was the means; more often, simply the reduction in grass cover brought about by grazing. The outcome was a dramatic recession in grasslands. Succulent deserts encroached from lower elevations; brush crept over hillsides; ponderosa forests swelled into dense dog-hair thickets and pinyon-juniper groves splashed across the landscape. Range deteriorated, site productivity diminished, and fuels built up to alarming quantities.

This regime is now changing once again as prescribed fire—natural and scheduled both—is introduced. In wilderness areas the abundant lightning fire of the region again burns. In the ponderosa pine belt across the Mogollon Rim, broadcast fire is applied during the late fall for fuel reduction. In brush-infested grasslands, broadcast fire is used, along with chemical supplements, to control the spread of woody plants, and on brushy watersheds similar techniques attempt to type conversion back to grass. On areas invaded by pinyon and juniper the trees are knocked down, left to dry, and then burned with the expectation that grasses will once again establish themselves.

Of the major vegetation types in southeastern Arizona, only the ponderosa pine/mixed-conifer forests and the grasslands have received much attention from researchers interested in wildfire. For the most part, neither the impact of wildfire on the evolution of present vegetation nor the role of fire suppression is understood. In fact, Hastings and Turner (1965) question whether wildfires were ever common in southeastern Arizona or that wildfires and fire suppression had affected vegetation changes since the 1880s. Still others, while recognizing the effects of fire suppression on the region's montane coniferous forests, do not acknowledge the effects of wildfire on the evergreen woodlands and desert grasslands, in spite of the fact that before the heavy overgrazing of the 1880s and 1890s both lightning and anthropogenic fires were common in both vegetation types (Bahre 1985; Baisan 1988).

Hastings and Turner (1965) question whether the Indians commonly set fire to the vegetation historically, especially to the grasslands. Dobyns (1981), Pyne (1982), and Bahre (1985), however, using information from early Spanish and Mexican records, contact ethnographies, and nineteenth-century newspaper accounts, present evidence that the Indians, especially the Apaches, set wildfires both before and after Anglo settlement. Indians were notoriously careless with fire, and the historical ubiquity of wildfires set by Indians in the grasslands, savannas, and forests of the New World is widely reported (Stewart 1956).

Hastings and Turner (1965) emphasize that fire drives employed for hunting were the only means by which the Indians communicated fire to the vegetation. Ignored are abandoned campfires, mescal roasting

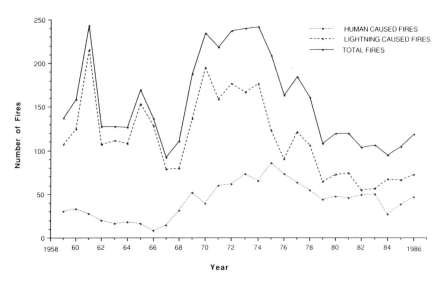

FIGURE 6.1. Lightning versus anthropogenic wildfires in the Coronado National Forest, 1959–1986. *Source*: John Turner, Coronado National Forest, November 1987.

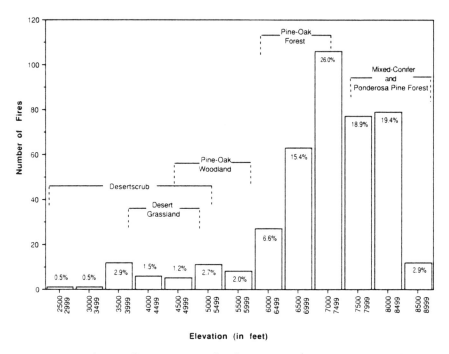

FIGURE 6.2. Lightning fire occurrence by elevation in the Rincon Mountains, 1937–1986. *Source*: Baisan 1988:30.

Table 6.1. Coronado National Forest Fire Statistics, 1959–1986

	Human-Caused		Lightning-Caused		Total	
Year	No. Fires	No. Acres	No. Fires	No. Acres	No. Fires	No. Acres
1959	30	1,072	107	478	137	1,550
1960	33	1,356	125	15,528	158	16,884
1961	28	791	215	14,657	243	15,448
1962	20	725	107	1,754	127	2,479
1963	16	234	111	216	127	450
1964	18	164	108	1,310	126	1,474
1965	16	642	153	4,946	169	5,588
1966	8	326	128	2,130	136	2,456
1967	14	4,352	78	833	92	5,185
1968	31	3,638	79	25,266	110	28,904
1969	52	4,463	136	4,914	188	9,377
1970	39	3,214	195	2,707	234	5,921
1971	60	1,680	158	3,525	218	5,205
1972	61	380	176	731	237	1,111
1973	73	829	166	659	239	1,488
1974	65	3,025	176	10,315	241	13,340
1975	85	1,344	123	982	208	2,326
1976	73	1,285	90	851	163	2,136
1977	63	9,433	121	255	184	9,688
1978	54	1,143	106	1,399	160	2,542
1979	44	3,330	64	2,142	108	5,472
1980	47	915	72	1,313	119	2,228
1981	45	112	74	1,525	119	1,637
1982	49	450	54	371	103	821
1983	50	2,742	56	383	106	3,125
1984	27	143	67	1,850	94	1,993
1985	38	1,848	66	8,788	104	10,636
1986	46	525	72	11,865	118	12,390
Total	1,185	50,161	3,183	121,693	4,368	171,754

SOURCE: John Turner, letter to the author, Nov. 1987.

fires (Brady n.d.; *Weekly Arizonian* June 9, 1859), smoke signals (Chalmers n.d.; *Arizona Weekly Citizen* July 5, 1873), and the setting of fires during warfare (*Arizona Daily Star* Mar. 24, 1882, June 15, 1887; Kellogg 1902a:505)—all of which just as easily led to wildfires. Pyne (1982:72) notes, too, that the "Apaches were said to burn off miles of mountain landscape in the 'delusion that the conflagration would bring rain.'"

Dobyns (1981:42), compiling a list of the Sonoran Desert Indian tribes who hunted with fire, details the effects of an Apache fire drive in the Aravaipa Canyon area in 1830. Given their land-use activities and attitudes toward fire, the Indians were surely more prone to ignite wildfires than were the nineteenth-century Anglo settlers and forest administrators who openly advocated suppression of wildfire. In any event, given the high incidence of lightning-caused fires, reliably documented by modern data, the relative importance of fires set by Indians is probably moot.

Overgrazing of rangelands has been a major means of fire exclusion in southeastern Arizona. Leopold (1924) reported that the primary cause for the increase of woody plants in southeastern Arizona early in the twentieth century was a decline in the incidence of wildfire due largely to overgrazing. According to him, the prevalence of charred stumps and fire scars on most of the older trees proved that wildfires were frequent before the overgrazing of the 1880s and 1890s. Even in the 1880s, however, wildfires were fairly common, especially in the grasslands. Clara Spalding Brown (1893:525), one of the first female residents of Tombstone, described a prairie fire along the road from Charleston to the Huachuca Mountains in the early 1880s:

> Then we came upon a prairie fire which we had observed raging in the distance. It was moving swiftly along our right, but as the grass on our left had been previously destroyed, it did not menace us with danger. It was a weird sight, not without beauty, as the remorseless line advanced, darting up vivid forks of flame, and presently became entangled with a whirlwind, a dense black cloud like a waterspout moving along in the track of the fire, and ever and anon revealing in its dusky heart tongues of flashing light, a startling contrast to their inky background. The burning grass, though brown and sear [sic], was full of nutrition, and was a great boon to cattle in a country where only the river-bottoms and certain moist valleys like the San Simeon [Simon] yield verdure.

Since establishment of the first National Forest Reserves in southeastern Arizona in 1902, the policy of Forest Service administrators until the 1980s has been one of almost total fire exclusion and suppression. Leopold (1924:9–10) points out that overgrazing was practiced by forest administrators in Arizona at the turn of the century to reduce fire hazards and promote the growth of trees. Arthur Noon, one of the first rangers in the Huachuca National Forest Reserve, notes that the Huachucas were so full of cattle and cattle trails that the trails served as good firebreaks (Winn n.d.).

NINETEENTH-CENTURY WILDFIRES

Most ecologists consider wildfire important to the evolution of coniferous forests, evergreen woodlands, and grasslands of southeastern Arizona, but until the 1980s few primary historical sources or fire-scar records were cited to document the occurrence and scale of nineteenth-century wildfires (Bahre 1985). This lack of primary research documentation may be one reason why some researchers have questioned the importance of wildfire in the evolution of the region's grasslands and evergreen woodlands.

However, local newspaper accounts of wildfires in southeastern Arizona between 1859 (the year the first newspaper in southeastern Arizona was published) and 1890 demonstrate that during that period (1) wildfires were much larger in areal extent, especially in the grasslands, than they are now; (2) the occurrence of large grassland fires seems to have declined after the 1880s, probably as a result of overgrazing; (3) the cessation of major grassland fires preceded the "brush invasion" of the late 1890s; (4) Indians, especially Apaches, set wildfires; (5) wildfire suppression was favored by early Anglo settlers; (6) wildfires occurred in all of the major vegetation types, including desertscrub; and (7) wildfires were fairly frequent (Bahre 1985).

Table 6.2 lists the major wildfires per year in southeastern Arizona reported in local newspapers (largely from Tucson and Tombstone) from 1859 to 1890 and indicates the major vegetation type(s) in which each fire occurred. The table offers conservative estimates of the number of fires because in some cases the same fire was reported many times (e.g., *Arizona Weekly Star* June 19, 1879; *Arizona Daily Star* July 30, 1879) or an unspecified number of fires were noted in several different mountain ranges (e.g., *Tucson Daily Record* June 4, 1880). Furthermore, the newspaper accounts do not always describe the vegetation type(s) in which the fires occurred. For example, many fires are noted only to have occurred in certain mountain ranges, and it is not clear whether they were in the ponderosa pine and mixed-conifer forests, in the evergreen woodlands, or in both. Because the ponderosa pine and mixed-conifer forest represents a little more than 1 percent of the region's vegetation cover and occupies the highest elevations, where lightning frequencies are greatest, it may be overrepresented in the table.

The wildfires reported here represent only a few of those that probably occurred; newspaper coverage was spotty (largely because publication was irregular) and many issues have been lost. The most complete

Table 6.2. Wildfire Frequency in the Major Vegetation Types of Southeastern Arizona, 1859–1890 (data from local newspapers*)

Year	Estimated No. of Wildfires	Month and Day Reported	Major Vegetation Type			
			Ponderosa Pine and Mixed-Conifer Forests	Evergreen Woodland	Grassland	Desert-scrub
1859	1	6/2	X	X	X	
1870	1	5/28			X	X
1874	2	6/13			X	
1876	3	7/1	X	X	X	X
1877	5	6/21	X	X	X	X
		6/23	X	X	X	
		6/30	X	X	X	
1878	1	10/26	X	X		
1879	4	5/22	X	X	X	
		6/19	X	X		
		6/26	X	X		
		8/1	X	X		
1880	10	5/1	X	X		
		5/8		X	X	X
		5/15	X	X		
		5/15	X	X		
		6/2	X	X		
		6/3			X	X(?)
		6/4	X	X		
		6/4	X	X	X	

*Complete references to the wildfires noted here can be found by matching the dates with the newspapers in the bibliography.

daily and monthly coverage is for Tucson and Tombstone, and then only after 1879; before 1879 coverage is extremely poor, with no newspapers for 1860–1869 and only limited coverage for 1870–1878. The poor coverage is due in part to the fact that before 1878 Tucson was the only large settlement in southeastern Arizona that had much newspaper coverage. In 1860 only 372 people were counted in the region that now encompasses all of Cochise County as well as Santa Cruz and Pima counties east of Tucson and the Santa Cruz River (U.S. Bureau of the Census 1864) (see Table 2.1). In 1880, only 1,783 people were listed in that region, 1,473 of whom lived in Tombstone and its mill towns (U.S. Bureau of the Census 1883b) (see Table 2.1). Newspapers for other

Table 6.2. (continued)

Year	Estimated No. of Wildfires	Month and Day Reported	Major Vegetation Type			
			Ponderosa Pine and Mixed-Conifer Forests	Evergreen Woodland	Grassland	Desert-scrub
1882	8	2/24	X	X		
		3/17			X	X
		4/3			X	
		4/16	X	X	X	
		6/27	X	X		
		7/19	X	X		
		9/27			X	
1883	7	6/2			X	X
		6/9		X		
		6/23	X	X		
		6/23	X	X		
		6/30		X	X	X
		6/30		X	X	
1886	3	4/7		X		
		5/12		X		
		7/7	X	X		
1887	4	5/17	X	X		
		6/9	X	X		
		6/14	X	X	X(?)	
		6/15	X	X		
		9/21	X	X		
1889	1	6/22	X	X		

towns in the region have been examined, but their publication was erratic and entire years are missing.

The largest number of fires was recorded for the mountains surrounding Tucson—the Santa Catalinas, Rincons, and Santa Ritas—and after 1878, the mountains to the east and west of Tombstone—the Dragoons, Huachucas, and Whetstones—mainly because those towns had the most complete newspaper coverage. Generally, only wildfires that could be seen from Tucson or Tombstone, or those that were extremely large or threatened personal property, were recorded. In almost every case, wildfires were given minor coverage.

For a more thorough perspective on the nature of the nineteenth-century fires, the following complete newspaper descriptions are included, in chronological order:

> Fire on the Mountain—All the past week great fires have been raging along the western slope of the Santa Rita Mountains, extending sometimes to the tallest peaks. At night the scene was grand—a vast illumination of the mighty hills—the fire in circles, in long lines, in scattered patches, and glowing in the distant horizon like the watch-fires of a great army.... The entire western slope of the mountains has been burned over, and the fires are now working over and around to the eastern side, making at night a strange and beautiful spectacle. (*Weekly Arizonian* [Tubac] June 2, 1859)

> Indian signal fires have blazed up during the week at various points in the mountains within sight of Tucson. (*Arizona Weekly Citizen* July 5, 1873)

> Fires have been raging south and southeast of here [Tucson] during the past week. Millions of acres of excellent grass land have been burned over but thanks to the abundance of our grazing lands we have plenty left. As soon as the rainy season sets in, which will be about the first of next month, the whole county will again be covered with green grass. (*Arizona Weekly Citizen* June 13, 1874)

> ... People are rapidly going into the Dragoon and other mountains on the abandoned reservation [Chiricahua Apache Reservation], and the outlook is generally encouraging. The only misfortune now out that way is the general burning of the country—whether by design or accident.... Fires have extended from the Santa Catarina [Catalina] Mountains near Tucson, through the canyons to the San Pedro Valley, and also from the east in like manner. The worst effect is the general destruction of timber in the canyons and mountains. People ought to be careful in setting fire to the grass in the dry seasons. (*Arizona Weekly Citizen* July 1, 1876)

> For the last month the country north, south and east of Tucson has been in a constant blaze. The grasses on the mesas, mountains and in the valleys, have been eaten up by the flames; during the last two days the fire has traveled over the Santa Catarina [Catalina] mountains and is burning now miles beyond. It has climbed almost to the summit of the Santa Rita, after devouring most of the pastures below, and bids fair to continue its course until the grass of the whole country has been licked up in flames. At first thought, these fires might appear to be of little or no harm to the country or people, but a moment's reflection convinces us of our error. We lay claim (and justly) to great and extensive grazing facilities in southern Arizona; and large herds of stock are constantly coming in to eat up these fine pastures; but let these grasses be burnt off two or three years in succession during the hot dry season and not only grass but all kinds of vegetation must disappear, the roots will be destroyed by the penetrating heat. The shrubbery and trees will be consumed in the same way, wherever the fire touches. The timber in the Santa Rita mountains now being destroyed, ought to have brought us in the future millions of feet of lumber; the small

fuel timber on the mesas, which ought to have supplied us for years to come is snatched away by the same element. (*Arizona Star* June 23, 1877)

It is a permanent misfortune to have the country burned over as it is in places within sight of Tucson, especially north of town. We are inclined to think the Papagos set out the fire northward to aid them in hunting, but whoever does it, does a great wrong. Fires have also been seen in the Santa Rita and Santa Catarina [Catalina] Mountains and also in the Rincon, where good timber must have been destroyed. In view of these fires, heavy rain can not fall too quickly. (*Arizona Weekly Citizen* June 30, 1877)

For the past week a fire has been burning in the pine timber that covers the higher portions of the Santa Catarina [Catalina] mountains. Daily the clouds of smoke hanging over the summit and the bright lights at night, show that great damage is being done. For over a month during the summer of last year similar fires could be seen surrounding the town, in a semi-circle, from the Tortillata [Tortolita] mountains on the north to the Santa Ritas on the south. Too little heed is given to these fires and the loss they occasion is rarely considered. It is hard, because of the lack of any data upon which to base calculations, to estimate even approximately the damage. The following however may serve to bring the matter home in somewhat tangible form. . . . It is an exceeding low estimate, because of the extent of the fires, to assume that one tree was consumed every three minutes, or twenty each hour during the month of last year referred to. This would make for the month 14,400 trees destroyed absolutely; in round numbers say 15,000 trees. (*Arizona Weekly Citizen* Oct. 26, 1878)

Fires have been raging in the mountains east and south of here [Tucson] for some time past, and much valuable timber and grass have been destroyed. We are not favored with such a superabundance of either of these articles that they should be maliciously or carelessly destroyed. There is a law with heavy penalties against setting fires of this kind; but it is practically a dead letter, owing to the difficulty of procuring a conviction under it. (*Arizona Weekly Star* May 22, 1879)

Florence. . . . Gen. Rice and party have arrived from the Santa Catarina [Catalina] mountains and report large tracts of country burned by Indians on the eastern slope of the mountains and a large quantity of valuable timber has been destroyed. (*Arizona Weekly Citizen* Aug. 1, 1879)

. . . There has been a fire raging in the Sierrita Mountains for the past week, which has burned the wood and grass from summit to base for some twenty miles in length and doing much damage. (*Arizona Weekly Citizen* May 8, 1880)

Forest fires are raging in the Whetstone and Huachuca mountains and also on the northside of the Santa Catarinas [Catalinas]. Acres of fine grass and timber are being destroyed. Ranchmen and prospectors are much annoyed by the carelessness of campers in allowing these fires to get underway. Grass through these mountains is now in its best condition and it is a pity to have it destroyed. (*Tucson Daily Record* May 15, 1880)

For the past two weeks fires have been burning on the Santa Catalina, Santa Rita, Pajarito and Oro Blanco mountains; during that time over 100 square miles have been burned over, destroying not only all grass but also all the trees and timber in the burned districts. The grass alone destroyed is an important item, but the value of the timber it is impossible to compute. These fires are nearly all caused by careless prospectors; we say careless because we cannot believe that anyone could be mischievous enough to set these fires. If these prospectors would only think for a moment and realize the immense damage they do, not only to their fellow prospectors and miners but the whole country. . . . Whatever the name Arizona will be derived from is immaterial, but this is true, and this is an "Arid zone" caused almost entirely by these carelessly or designedly set fires. Fires have destroyed and are destroying our trees and timber, and where they are more existent rains and moisture will not be attracted, and where moisture does not exist herbage cannot. Is it then surprising that this is an "Arid zone", and is it not plausible that from this fact the name Arizona was derived. The name will probably exist as long as time lasts, but the cause which gave birth can be removed, by care on one hand and the rigid application of the laws on the other. (*Tucson Daily Record* June 4, 1880)

. . . the grass over areas that were burned over this season is now knee high and everything looks as fresh as spring time in this locality [Patagonia]. . . . (*Arizona Daily Star* Sept. 2, 1880)

The sky in the west was brilliantly illuminated last night, which indicates a great fire raging far beyond this place [Tucson]. Both Maricopa and Casa Grande telegraphed here as to the cause. The light from the clouds was reflected with brilliancy on the city. (*Arizona Daily Star* Mar. 17, 1882)

Prairie and wood fires have been raging in southern Arizona and western New Mexico recently. The territory burned over is reported to cover forty miles square, and the damage done is immense. The origin of the fire is attributed to the Indians. (*Arizona Daily Star* Apr. 16, 1882)

Immense forest fires are still prevailing in some parts of western New Mexico and southern Arizona. They are believed to have been set out by Indians. Next to the pleasures of killing, burning appears to be the favorite amusement of the savages. (*Arizona Daily Star* May 21, 1882)

Fires raging near the San Pedro beyond the Cottonwoods. Large area burned to the west of Tucson. (*Arizona Weekly Citizen* June 2, 1883)

Fires can be discovered in the mountains many miles south of us, any night. (*Arizona Daily Star* June 23, 1883)

Extensive fires are raging in the north end of the Whetstone Mountains. (*Arizona Weekly Citizen* June 30, 1883)

The Patagonia mountains are on fire and the country between the Patagonia and the Huachucas, a distance of twenty miles, is covered with smoke. The tall grass and the pine timber is burning furiously, the noise being like a rushing storm. The heat is so great that one cannot approach within a

distance of it. An area of country about five miles square is now burning. It
was set on fire by the Indians, the day they murdered poor Grace. (*Arizona
Daily Star* June 14, 1887)

A large fire is raging in the Santa Catalina Mountains. A party who has been
at the scene states that it has gone over an area of about ten miles square
and that it is making its way toward the valley on this side [Tucson].
(*Arizona Daily Star* June 22, 1889)

The newspaper accounts clearly demonstrate that wildfires occurred in nearly all of the major vegetation types of southeastern Arizona and that they were generally large in the nineteenth century. By comparison, since about 1930 the largest fire in the grassland covered only 2,000 acres (Bahre 1977), the largest fire in the ponderosa pine and mixed-conifer forests and evergreen woodlands was the 9,800-acre Huachuca burn in June 1977 (Gehlbach 1981:184), and the largest fire in desertscrub was the 28,000-acre Granite burn southeast of Florence in July 1979 (McLaughlin and Bowers 1982). It is also noteworthy that the nineteenth-century newspaper accounts rarely identify lightning as the cause of wildfire, although there is good cause to doubt the newspapers' allegations that most of the fires were set by Indians, campers, and miners.

According to the nineteenth-century newspaper descriptions, wildfires were roughly twice as frequent in the ponderosa pine and mixed-conifer forests and evergreen woodland as in grassland, and about three times as frequent in grassland as in desertscrub (Table 6.2). This appears to match the fire-frequency data by elevation for the Rincon Mountains (Figure 6.2). Except for the grassland fire between the Patagonia and Huachuca mountains in 1887 (it is not clear whether that fire burned the grassland of the San Rafael Valley or the oak woodland-grassland mosaic of the Canelo Hills), major wildfires in the grasslands apparently declined after 1883, a year that corresponds to the beginnings of heavy overgrazing of grasslands (Wagoner 1961; Bahre and Bradbury 1978). Shortly after the drought of 1891–1893, the "brush invasion" of the grasslands supposedly began. A map of southern Arizona drawn by H. M. Robert in 1869 shows extensive grasslands all along the west side of the Dragoon Mountains and the east side of the Huachuca Mountains, in areas now heavily covered by mesquite and acacia (Figure 6.3).

IMPACTS OF WILDFIRE AND FIRE EXCLUSION

Although prescribed burning has been applied to the ponderosa pine and mixed-conifer forests, pinyon-juniper woodlands, and chaparral of

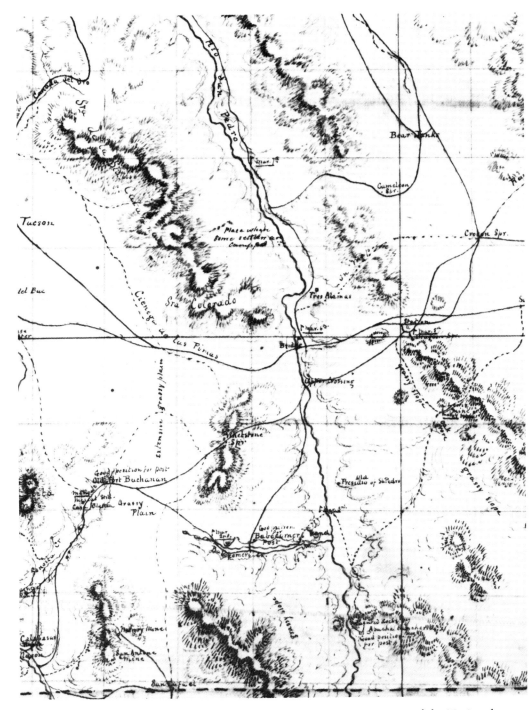

FIGURE 6.3. H. M. Robert's map of southeastern Arizona in 1869. Courtesy of the National Archives, Cartographic and Architectural Branch, Washington, D.C.

Arizona (Arnold et al. 1964; Zwolinski and Ehrenreich 1968; Pase and Granfelt 1977; Carmichael et al. 1978; Wright and Bailey 1982), and natural wildfires are now permitted to burn in wilderness areas, fire prescriptions are at present being advocated by some for the grasslands (Wright 1980). On the other hand, evergreen woodland, which makes up about 19 percent of the vegetation cover of the study area, has received little attention from fire ecologists, even though most oaks, especially the live oaks, resprout vigorously after fire (Phillips 1912; Longhurst 1956; Pase and Granfelt 1977; Carmichael et al. 1978; Griffin 1980; Wright and Bailey 1982; Bahre and Hutchinson 1985). Likewise, the two most extensive vegetation types—Sonoran and Chihuahuan desertscrub—have until recently attracted little attention (Humphrey 1974; Rogers and Steele 1980; McLaughlin and Bowers 1982; Brown and Minnich 1986; Rogers 1986; Rogers and Vint 1987).

The primary objectives of prescribed burning have been (1) to improve livestock forage by eliminating so-called undesirable plants, reducing litter, and stimulating growth of desirable forage plants; (2) to improve wildlife habitat; (3) to reduce fuel buildup and thereby lower the incidence of catastrophic fires that have plagued forests since the initiation of fire suppression policies; and (4) to restore the natural role of wildfire in the vegetation.

Although it is now widely recognized that wildfire exclusion has increased the hazard of catastrophic fires due to fuel accumulation, especially in the montane coniferous forests of the American Southwest (Vogl 1971a, 1971b; Dodge 1972; Wright and Bailey 1982; Pyne 1982), only during the 1980s have "natural prescribed" lightning fires been allowed to burn in some National Forest lands in southeastern Arizona. Since the Yellowstone fires of 1988, however, that policy has (at least temporarily) been changed back to one of fire exclusion, except for some "planned prescribed burning."

Weaver (1968, 1974), who did much of the pioneer work on fire and ponderosa pine in the United States, especially in the Mogollon Rim country, believes that ponderosa pine is best adapted to a regime of frequent periodic burning and that wildfires were the primary ecological influence in molding the forest as the first white man found it. The dendrochronological record of fire scars in the ponderosa pine forests of the American Southwest supports his thesis. According to Swetnam (in press), ponderosa forests of the Mogollon Rim experienced surface fires at an average interval of four or five years before the arrival of the white man with his logging operations, livestock, and fire suppression policies. Swetnam (ibid.) states, however, that the pre-Anglo mean fire interval in the ponderosa pine forests of Arizona and New Mexico varied from two to ten years, while the mean fire interval between 1657 and 1893 for the Rincon Mountains east of Tucson was 6.1 years (Baisan 1988:63). After 1893 this "more-or-less regular pattern of fire occurrence ceased" (ibid.).

Much less is known about fire intervals in the mixed-conifer forest, although the fire frequency is lower because this forest type is more mesic (Weaver 1951a; Ahlstrand 1980; Dieterich 1983). The interval recorded by Baisan (1988:63) for the mixed-conifer forest in the Rincon Mountains was 9.9 years. Surprisingly, no pre-twentieth-century fire frequency data have been compiled for the ponderosa pine and mixed-conifer forests of the other mountain ranges in southeastern Arizona. Besides leading to the buildup of fuel and more intense, devastating crown fires in the ponderosa pine and mixed-conifer forests, fire exclusion has affected other aspects of forest ecology in southeastern Arizona. For example, Sawyer and Kinraide (1980:239) maintain that Douglas fir (*Pseudotsuga menziesii*) has increased greatly in the Chiricahua Mountains since the cessation of fires that ravaged the range nearly a century before, and Russell (1982) reports that fire exclusion has been the primary cause for the desiccation and tree invasion of the high mountain meadows in the same range.

The vegetation type(s) receiving the most attention from researchers interested in fire ecology in southeastern Arizona are the plains and desert grasslands; these have undergone major brush and scrubby tree increases since the 1891–1893 drought. The increase in woody plants in southeastern Arizona, however, probably does not represent an invasion, because woody plants were always present in the grasslands, albeit in much lower numbers than at present.

In his discussion of the invasion of the desert and plains grasslands of the Southwest, Harris (1966:416) speculates:

> Definite knowledge of the ecological status of the desert and plains grasslands at the time of European settlement awaits fuller investigation of the post-Pleistocene history of vegetation in the Southwest, by pollen analysis and other means; but meanwhile it is reasonable to suppose that they were secondary or "disclimax" communities derived from a former cover of woody vegetation in which mesquite may well have been an important element. If so, then the spread of mesquite should be seen not as an invasion of natural, primary grassland but as part of the reoccupation by woody plants of areas in which secondary grass communities had attained dominance as a result of human interference.

Of course, lightning and anthropogenic fires may have caused or maintained the dominance of grasses before the period of Anglo fire exclusion. But if lightning fires were always a part of the ecosystem, how could one reasonably speak of a former cover of woody vegetation? While there appears to be a close relationship between fire and the maintenance of relatively brush-free grasslands, that relationship is not simple, and the effects of fire suppression on rangelands remain unclear (Wright 1980).

Hastings and Turner (1965:38–40), examining twenty-two journals of early travelers through southeastern Arizona, do not find the "slightest historical justification for applying the fire hypothesis to the desert grasslands of the region." These journals, however, cover only twelve trips between 1846 and 1858 that were of a few days' duration along heavily used and overgrazed trails through desertscrub and desert grassland. At any rate, because Hastings and Turner find no evidence that wildfires occurred at the requisite scale or frequency to explain the region's brush-free grasslands before the 1890s, they conclude that the shrub increase and grass decrease in the rangelands were primarily the result of climatic change, that is, a gradual trend toward higher temperatures and lower precipitation since 1870. However, a majority of knowledgeable researchers feel that there is no evidence that the climate has changed enough since 1870 to be a major factor in the rapid increase of woody shrubs and trees (Paulsen 1956; Humphrey 1958, 1987; Buffington and Herbel 1965; Cooke and Reeves 1976). Furthermore, droughts have been found to be just as hard on young mesquites as they are on grasses (Bogusch 1952; Carter 1964; Archer et al. 1988).

According to Wright (1980), the change from grass to brush since shortly before 1900 was due to a combination of factors, mostly related to grazing: (1) reduced grass fuel, which has lessened the ability of the range to carry fire and has lowered fire temperatures—and, hence, has reduced the ability of fires to kill shrubs and trees; (2) decreased activity of jackrabbits and wood rats, which depend heavily on mesquite seedlings for food; (3) increased erosion, which helps cover and irrigate mesquite seed; (4) increased seed dispersal by livestock; (5) increased seed source, as more trees are able to mature because they are not killed by fire; and (6) reduced competing stands of perennial grasses, which, when healthy and dense, can keep woody trees and shrubs in check.

Wright (1980) further notes that except for black grama (*Bouteloua eriopoda*), which takes from three to eight years to recover from fire, most grasses recover in one to three years, and that fire in the dry season from April through June, especially following one to two years of above-average summer precipitation when herbaceous fuel has built up, will kill most of the burroweed (*Haplopappus tenuisectus*), cactus, broom snakeweed (*Gutierrezia sarothrae*), creosote bush (*Larrea tridentata*), and young mesquite plants while leaving most desirable forage plants unharmed. His findings are similar to those of Humphrey (1958), Cable (1967), Bock et al. (1976), and Bock and Bock (1978), who demonstrate that the native plants and animals of the plains and desert grasslands of southeastern Arizona evolved under conditions of frequent fires and were left unharmed by their occurrence.

There is a tremendous amount of published research on the effects of fire in controlling woody shrubs and trees in the desert grasslands. This research is comprehensively reviewed by Humphrey (1958) and Wright (1980). Here are some of the more significant results:

1. Vigorous perennial grasses compete strongly with mesquite seedlings (Martin 1975; Wright et al. 1976). Martin (1975) finds that competition from healthy grass can reduce the number of mesquite plants by 94 percent, while Glendening and Paulsen (1955) note that sixteen times as many mesquite seedlings are established in bare areas as in vigorous stands of perennial grasses.

2. Mesquite seedlings are particularly vulnerable to fire during the dry season in late spring and early summer, especially after one to two years of above-normal rainfall when there is a substantial buildup of fine fuel (Glendening and Paulsen 1955; Cable 1972; Wright 1980). Cable (1961) finds that one grass fire killed 67 percent of six- to twelve-month-old mesquite seedlings. Reynolds and Bohning (1956) report that a grass fire killed 40 percent of mesquite trees less than two inches in diameter, and Glendening and Paulsen (1955) find that a June fire on the Santa Rita Experimental Range killed 60 percent of mesquite trees less than half an inch in diameter, 20 percent of trees one to two inches in diameter, and 11 percent of trees over five inches in diameter. Cable (1961, 1965, 1967, 1972) and White (1969) report that mesquite kill decreased as stem diameter increased, that large mature mesquite were not killed by grass fire (which might indicate the imbalance caused by historic fire exclusion), and that the effect of fire on mesquite depends on the size of the mesquite and the amount of fine fuel available for burning. Tschirley and Martin (1961) report that the mortality of burroweed in June fires was about twice as high as in fires in late fall or early spring. Humphrey (1949) notes that June fires were about 100 percent effective in killing burroweed. (Note: The highest incidence of wildfires is in June and early July.)

3. Before the 1880s and 1890s, when wildfires were more frequent than they are today, it is probable that few mesquite seedlings in the up-

land desert grassland survived long enough to become mature trees and set seed (Humphrey 1974:399).
4. Periodic burning can maintain a grassland aspect if the intensity and frequency of grazing allow enough dry herbage for an effective fire that can kill most scrubby trees and brush (Martin 1983:604).

Today, some range managers advocate prescribed burning in the grasslands to destroy brush (Martin 1975; Wright and Bailey 1982; Cave and Patten 1984). However, because of overgrazing, in many areas the fine fuel load is too low to sustain fire. Furthermore, many mesquite trees are now too large to be killed by grass fires.

Wildfires also occurred in desertscrub communities during the nineteenth century; in fact, they may have been more common than previously thought. Although desert lightning storms are frequent during the summer, it is generally believed that the production of annual and perennial herbs is usually too low to provide a fuel load capable of sustaining wildfire (Humphrey 1963, 1974). Consequently, Sonoran and Chihuahuan desertscrub communities in Arizona have been little studied by fire ecologists. Occasionally, after consecutive years of greater-than-normal precipitation, the fuel load may be such that, if the land is not overgrazed, fires can occur. This happened on the Granite burn south of Florence in June 1979 (McLaughlin and Bowers 1982).

A recent increase in the number of fires in desertscrub has renewed interest among fire ecologists in this vegetation type in southern Arizona (Rogers and Steele 1980; McLaughlin and Bowers 1982; Cave 1982; Cave and Patten 1984; Rogers and Vint 1987). Rogers and Steele (1980) believe that wildfires may be more prevalent now in Sonoran desertscrub because introduced annuals have increased fuel supplies. However, several of the exotic species that Rogers (1986) considers to be prime fuels, such as *Erodium cicutarium, Bromus rubens,* and *Schismus barbatus,* have been widespread in Arizona since the middle of the nineteenth century, and the first two of these are major forage plants. Furthermore, according to the historic record, there may have been more fires in desertscrub in the nineteenth century than recently, presumably because there was less grazing then. Whatever the incidence of desertscrub wildfires, most of the desert species appear to have a low tolerance to burning, and adaptations to fire in this vegetation type have not been strongly developed (Rogers and Steele 1980; McLaughlin and Bowers 1982; Rogers 1986).

Humphrey (1974:381–382) notes that tobosa and sacaton flats in Sonoran and Chihuahuan desertscrub may have been burned regularly in the past, but are now being invaded by woody plants because of fire exclusion. Indeed, many ranchers who ran cattle forty or more years ago in southern Arizona report that they used to burn areas of tobosa (*Hilaria mutica*) and sacaton (*Sporobolus* spp.) to green up the grasses for forage before state and federal agencies put a stop to this practice.

Anglo-Americans changed the frequency and ecological role of wildfire in southeastern Arizona by overgrazing and purposeful fire exclusion. And even though the role of wildfire in the evolution of the vegetation of southeastern Arizona has been little studied, especially in the evergreen woodland and desertscrub communities, it is clear that wildfire exclusion since the advent of major Anglo-American settlement has had a major impact on the region's grasslands and ponderosa pine forests.

7 Fuelwood Cutting

Cordwood, scarce in the predominantly desertscrub and grassland vegetation of southeastern Arizona, has historically been the most important, and often the only, source of fuel (Bahre 1984; Bahre and Hutchinson 1985). Yet few students of vegetation change in Arizona realize how much fuelwood was cut from desert woodlands for nineteenth-century mining operations or how much regeneration has occurred in those woodlands since the early 1940s, when cordwood ceased being the major fuel for domestic heating and cooking.

According to the record, fuelwood cutting once had a major impact on the riparian forests, mesquite thickets, and evergreen woodlands near most of southeastern Arizona's major cities and mining centers. For example, in 1892, because fuelwood was so scarce near Tucson, woodcutters had to "go as far out as 20 or 30 miles" and even then brought in "roots and stumps, dug and cut up into stove size" (*Arizona Daily Star* Dec. 20, 1892). By 1905, reportedly "every tree over seven inches in diameter had been cut and used for fuel within a ten-mile radius of Tucson" (Harrison 1972:103). At Morenci in 1890, "Firewood had disappeared from above ground on the hills around the town, but the resourceful Mexican was still supplying his individual needs by digging for it" (Barr 1940:6). At Bisbee in 1882, the stripping of the hills of all timber within miles of town led to flooding (*Bisbee Review* Aug. 8, 1923). In 1884 it was noted that "timber depredations in southern Arizona are becoming so extensive that there is just cause for alarm. Even the palo verde trees are being stripped from the mesa lands" (*Arizona Daily Star* Mar. 7, 1884). That by 1882 cordwood had become scarce in southeastern Arizona was reported in the *Arizona Daily Star* (Mar. 1, 1882):

> The fuel question is one of the most important to the mining industry. The wood supply, as a matter of course, is fast decreasing. . . . Other sources must be sought. The coal fields is [sic] the only resort, and as those of San

Carlos are almost inaccessible at present, we must look beyond the Territory. The *Epitaph* says: There is soon to be a change in the fuel used at our mines and mills. Wood is to be superceded [sic] by coal from New Mexico, which it is believed can be laid down for $12 per ton, or at the same cost as a cord of wood.... At the present rate of consumption, if continued as the staple fuel, [wood] would become an exceeding dear commodity. As it is, that supplied to the mines comes mostly from the eastern slope of the Dragoons, from twelve to fifteen miles distant, and costs $12 per cord. That supplied for domestic purposes comes equally high, and this cold weather eats large holes into people's income....

Except for the English and Colorado coke used in the blast furnaces, cordwood was the major fuel of the mines in southeastern Arizona until the 1890s. Wood was burned under the boilers of the steam engines at virtually every step in mining—to run stamps, pumps, hoists, ore crushers, dryers, amalgamation pans, settlers, and converters; to roast ores; and to retort amalgam. Wood fueled every stream engine in the area, from those running trains to those making ice, and it also met all heating and cooking needs. Trees were cut from the ponderosa pine and mixed-conifer forests in the Huachuca, Santa Rita, Chiricahua, Pinaleño, Santa Teresa, and Santa Catalina mountains for construction and mine timber. In addition, miners and ranchers cut trees to make fenceposts, for which they favored juniper, mesquite, and desert willow (*Chilopsis linearis*); to make wood corrals; to kiln firebricks, especially willow and cottonwood; and to feed livestock (with the leaves of willow and the bark of cottonwood) (*Arizona Daily Star* Oct. 24, 1880; Bahre and Hutchinson 1985:179).

Charcoal, made mostly from mesquite and oak, was used for several purposes throughout southeastern Arizona, from blowing in smelting furnaces to heating laundry irons. Before the late 1870s it was widely used in the old Mexican adobe blast furnaces at such early mining camps as Clifton-Morenci, Globe, Superior, Bisbee, Harshaw, Mowry, Sopori, and Heintzelman; after 1880, the widely used water-jacketed blast furnaces used charcoal only to ignite coke. In some instances, coke had to be freighted long distances from rail stations. For example, the Detroit Mining Company at Clifton Morenci freighted its coke from Willcox, and the Copper Queen at Bisbee freighted its coke first from Benson and then from Fairbank until the railroad arrived in Bisbee. Most of the early mines in southeastern Arizona relied on animal power for hoisting and crushing ores and, although stamp mills powered by wood-burning steam engines first appeared in Arizona in the 1860s, it was not until the late 1870s that steam power became dominant (Bahre 1984:103–104).

Besides Tombstone and Bisbee, other mining centers in southeastern Arizona were at Dragoon, Pearce, Harshaw, Silver Bell, Duquesne, Washington Camp, Evansville, Courtland, Greaterville, Turquoise, Gleeson, Dos Cabezas, Helvetia, Middlemarch, Black Diamond, Paradise, Russelville, Rosemont, Johnson, Galeyville, Oracle, Mammoth, Total Wreck, Hilltop, and Winkelman (see Figure 1.1). Smelters were constructed at these centers as well as at Tucson, Lochiel, Hereford, Patagonia, Nogales, and Benson. Much of the cordwood for the Globe, Clifton-Morenci, and Silver King operations came from southeastern Arizona.

For the most part, the mining industry used mesquite, oak, pinyon, juniper, and some shrubs for fuel. Of these, oak and mesquite have the highest caloric values, but the extra labor needed for splitting and seasoning them sometimes offsets their better quality. Photographs of woodpiles next to hoisting works, pumps, stamp mills, and smelters show what appears to be a preponderance of these two species (see Figure 7.1). The mines demanded seasoned wood six inches or more in diameter. However, where fuel was scarce, even the smallest trees near towns and mining settlements were cut.[1] Juniper, oak, and mesquite were the primary heating and cooking fuels; also used were pine and shrubs such as *Arctostaphylos* spp., *Rhus* spp., *Cercocarpus* spp., *Garrya* spp., and *Ceanothus* spp. Juniper was particularly sought, because after seasoning it was easy to ignite and had a pleasing fragrance.

Local contractors, employing Mexican woodcutters, supplied most of the cordwood. From all available records, there must have been many woodcutters and woodcutters' camps, especially in the evergreen woodlands and mesquite thickets near the major towns and mining camps.[2] However, most contemporary writers took fuelwood cutting so much for granted that they said little about it. One exception was the *Tombstone Daily Nugget* (Feb. 5, 1882), which asked the citizens of Tombstone not to forget the woodcutters' importance to the economic survival of the city:

> Elsewhere we print what we consider authentic information as to a raid made on Friday by Indians [Apaches] in the Dragoon Mountains. Few of our people appreciate the amount of work done in the wood camps along the base of the mountain, the number of men who are exposed to danger should the raids continue, and how necessary for our mines that these men should be protected.

Although total fuelwood consumption for mining peaked during the 1890s, some mines continued to use fuelwood until shortly after the turn of the century. After 1890, however, most fuelwood was cut for

FIGURE 7.1. Ricks of cordwood at the Contention Mill, Contention, 1882. Photograph by C. E. Watkins, courtesy of the Bancroft Library.

domestic heating and cooking. Even as late as 1940, 44 percent of the occupied dwellings in Arizona used fuelwood for heating and cooking (U.S. Bureau of the Census 1943:110–114). Fuelwood consumption declined in southeastern Arizona after 1940, but since 1973 the demand for residential fuelwood has risen dramatically.

THE TOMBSTONE "WOODSHED"

For a better perspective on the site-specific effects of historic fuelwood cutting in southeastern Arizona, I have included a case study on the historic mining town of Tombstone and its "woodshed," the area within a twenty-five-mile radius of the settlement that has served as Tombstone's source of fuelwood since the 1870s.[3]

Tombstone, established in 1878, was one of several major mining centers in southeastern Arizona that relied heavily on fuelwood during the nineteenth century. Today it is a small community of about 1,700 people. The town's present appearance belies the fact that it was once

Arizona's leading silver producer and possibly its second largest city (15,000 in 1885 reported by McClintock [1916:vol. 2,582]). During the Tombstone bonanza (1879–1886) $19.4 million in silver—about $232 million at today's prices—was taken from the district's mines (Tenney n.d.). In Tombstone's heyday there were at least 50 different mines, 12 steam hoists, and 150 stamps at 7 mill sites (Blake 1902; Spude 1979). Because the silver strike occurred in desertscrub with scarce surface water, the stamp mills were located eight to nine miles west along the San Pedro River, the only perennial stream in the area, at Millville-Charleston, Boston (Emery City), Grand Central, Sunset (Headcenter), and Contention (Brown 1881; *Tombstone Daily Nugget* Feb. 18, 1882; Howe n.d.). Cordwood for the boilers of the engines that powered the mine equipment and for domestic heating and cooking was cut from the mesquite thickets along the San Pedro River and Babocomari Creek, and from the evergreen woodlands of the Huachuca and Whetstone mountains to the west and the Dragoon Mountains to the east.

The Tombstone woodshed covers about 2,000 square miles (about 31 percent of Cochise County and small parts of Pima and Santa Cruz counties) and encompasses nearly all of the major historical source areas of Tombstone's fuelwood (Bahre and Hutchinson 1985:179). There were eight major fuelwood-cutting areas in the Dragoon Mountains, and nearly every major canyon facing Tombstone in the Huachuca and Whetstone mountains was important. Sawmills for mining and construction timber were located in the ponderosa pine forests of the Huachuca Mountains, although most of Tombstone's sawtimber came from the Chiricahua Mountains, some fifty miles to the east.

Desertscrub and desert grassland cover about 84 percent of the Tombstone woodshed. Fuelwood is generally sparse in those communities, although mesquite is frequently dense along washes and in riparian areas along the San Pedro River and Babocomari Creek. Evergreen woodland covers about 15 percent of the woodshed; ponderosa pine and mixed-conifer forests cover only 1 percent (the high elevations of the Huachuca Mountains).

Coronado National Forest researchers have calculated the following average cordages per acre for the evergreen woodlands in the Tombstone woodshed: 4.41 cords in the Huachuca Mountains (where pure oak woodland averages only 1.29 cords) and 1.44 cords in the Dragoon Mountains. No fuelwood inventory has been conducted for the Whetstone Mountains. The low cordage estimate for the Dragoon Mountains may reflect past overcutting. With the possible exception of the Mule

Mountains near Bisbee, more fuelwood was removed from the Dragoon Mountains than from any other mountain range in southeastern Arizona before 1940. The Forest Service calculated cordage figures similar to those above for other evergreen woodlands in Cochise County: 2.73 cords per acre in the Peloncillo Mountains and 3.75 cords per acre in the Chiricahua Mountains.[4] According to these Forest Service figures, the average volume of oak and juniper in the Tombstone woodshed is about three cords per acre, or a total of about 587,000 cords. Not included in these cordage figures, however, is the considerable volume of mesquite, mostly on non-Forest Service lands, the total cordage of which probably exceeds that of oak and juniper. Annual growth rates or rates of accumulation on merchantable boles were estimated at only 1 percent; considering the rapid recovery of chained and burned oaks in the San Rafael Valley, this percentage seems conservative.

FUELWOOD CONSUMPTION IN THE TOMBSTONE WOODSHED

An estimated 31,000 cords of fuelwood were consumed for heating and cooking in Tombstone during the silver bonanza (1879–1886) (Bahre and Hutchinson 1985:182). This amount was derived by multiplying the population of the Tombstone District for each year of the bonanza by one cord of fuelwood per person annually (see Table 7.1) (or four cords per family for each occupied dwelling). This conservative estimate was derived from the average number of cords of fuelwood reportedly used for heating and cooking per year today by Indian and Mexican families in southern Arizona and northern Sonora.

During this same period, Bahre and Hutchinson estimate that Tombstone stamp mills consumed 47,260 cords of wood (1985:181). This amount was derived from the production charts in Tenney (n.d.) and the *Tombstone Daily Nugget* (Dec. 31, 1881), which indicate that 394,000 tons of ore were milled in the Tombstone District during the bonanza. The cordage consumed by the mills was estimated by multiplying the total tonnage of ore processed by 0.12 cord, the average amount of fuelwood reportedly required to mill one ton of ore (see Table 7.2).[5] The amounts of wood used to fuel the steam-powered hoists and pumps, and to roast ores and retort amalgam, are not reported. However, the fuelwood used for these purposes may have exceeded that required by the stamp mills. For example, Barr (1940) notes that one boiler for a pump on the San Francisco River near Clifton-Morenci consumed an average of four cords per day between 1880 and 1901 (about 30,000 cords in 21 years of operation), and Austin (1883) observes that four

Table 7.1. Domestic Fuelwood Consumption for the Tombstone Mining District During the Tombstone Silver Bonanza (June 1879–Dec. 1886)

Years	Fuelwood Consumed (cords)*
1879	1,000
1880	3,000
1881	6,000
1882	6,000
1883	7,000
1884	4,000
1885	2,000
1886	2,000
Total	31,000

SOURCE: Bahre and Hutchinson 1985.
*Based on reported population in censuses and newspapers, multiplied by one cord per person annually.

tubular boilers at the Harshaw silver mine required sixteen cords of wood per day.

Estimates of fuelwood consumption in the Tombstone woodshed after 1890 were taken from Cochise County data in the Sixteenth Census of Housing in 1940 and from the *Historical Statistics of the United States* (U.S. Bureau of the Census 1975:540). In 1940 fuelwood was used for cooking in 39 percent of Cochise County's 9,190 dwellings and for heating in 31 percent.

Based on Cochise County census data and the assumption that annual fuelwood consumption in each dwelling was four cords, Bahre and Hutchinson (1985:182–183) estimate that between 729,000 and 909,000 cords of fuelwood were consumed for domestic purposes from 1890 to 1940; about half that amount may have been harvested in the Tombstone woodshed. When combined with the estimates of fuelwood consumption for domestic purposes (Table 7.1) within the Tombstone woodshed during the bonanza, the total is between 396,000 and 486,000 cords.

Bahre and Hutchinson (1985:186) conclude that at least 443,000 cords of fuelwood were removed from the evergreen woodlands, mesquite thickets, and riparian forests of the Tombstone woodshed over

Table 7.2. Estimated Silver Ore Production and Stamp Mill Fuelwood Consumption During the Tombstone Silver Bonanza (June 1879–Dec. 1886)

Years	Production (dollars)	Silver Price (dollars/ounce)	Ore Processed (tons)*	Fuelwood Consumed (cords)†
1879–81	—	—	119,370	14,324
1882	5,202,876	1.14	115,629	13,876
1883	3,030,263	1.11	69,165	8,300
1884	1,380,788	1.11	31,516	3,782
1885	1,320,925	1.07	31,278	3,753
1886	1,050,000	.99	26,871	3,225
Total			393,829	47,260

SOURCES: *Tombstone Daily Nugget* Dec. 31, 1881; Tenney n.d.:19; Bahre and Hutchinson 1985.
 *The tons of ore processed from June 1879 through December 1881 were reported in the *Tombstone Daily Nugget* Dec. 31, 1881. The annual total for each year after that was estimated from Tenney's production figures, using Blake's average assay of 39.47 ounces of silver per ton of ore (1902:71):

$$\text{tons of ore processed} = \frac{\frac{\text{production (dollars)}}{\text{price per ounce (dollars)}}}{\text{average assay of 39.47 ounces of silver per ton of ore}}$$

†Tons of ore processed multiplied by 0.12 cord.

120 years. This figure does not include the fuelwood cut from the woodshed for other mining centers in southeastern Arizona, such as Bisbee and Pearce, for domestic heating and cooking for the towns and cities other than Tombstone and its mill towns before 1890, for mining processes other than milling, or for the bulk of the fuelwood cut since 1940. Most likely, however, the amount cut was twice this figure. In other words, the amount of cordwood consumed between 1878 and 1940 alone was more than the total cordage presently reported by the Forest Service for the evergreen woodlands of the Tombstone woodshed.

The consumption of cordwood by other mining centers in southeastern Arizona and northeastern Sonora, Mexico, was as great as in Tombstone. For example, Barr (1940) observes that the fuelwood consumed by the Morenci copper mine represented about 10 percent of the total cost of operation and that in 1885 between 300 and 500 cords were used every month. The Copper Queen at Bisbee burned 3,554 cords under its boilers in 1887 (*Bisbee Review* Aug. 8, 1923). The *Arizona Weekly*

Enterprise (Mar. 17, 1888) reported the following annual consumption of cordwood at the Silver King, just north of Florence:

> ... the consumption of fuel at the mine was 3,528 cords of wood (3,249 cords at the main shaft and 279 cords at Shaft No 2).... There are now on hand 571 cords at the main shaft and 95 cords at Shaft No 2.... The fuel consumed at the mill was 7,270 cords of wood of which 5,070 cords were for steam. There are now 654 cords on hand at the mill yard....

At Cananea and Nacozari, some thirty and sixty miles, respectively, south of the international boundary in Sonora, Mexico, wood-burning reverberatory furnaces were used for smelting. Unlike blast furnaces, which relied on coke, these furnaces were large consumers of fuelwood. According to Langton (1904:750), the cost of cordwood in 1901 had increased so much as a result of the exhaustion of wood supplies near Nacozari that only the completion of the railroad from Douglas would allow economical smelting. Parker (1979:79) observes that in 1902 "the countryside [near Nacozari] for 15 to 25 miles in all directions was entirely denuded of trees." Similar clearing of the woodlands was also reported around Cananea (Bahre 1984:107).

Since 1973, when petroleum costs rose, the demand for residential firewood, especially from Coronado National Forest lands, has increased tenfold. John Turner of the Coronado National Forest (interview, Mar. 1982) estimates that about 8,700 cords of "legal" fuelwood were cut from Forest Service lands in the Tombstone woodshed between 1971 and 1981. However, there are no data on how much wood, especially mesquite, was cut from private, state, or Bureau of Land Management lands. The demand for fuelwood from the Tombstone woodshed currently is rising rapidly, especially for residential heating in Tucson, Bisbee, Douglas, Sierra Vista, and Benson. The demand for fuelwood in Tucson alone from 1990 to 2000 is estimated at 195,925 cords (Ffolliot et al. 1979:29).

IMPACT OF FUELWOOD CUTTING ON THE TOMBSTONE WOODSHED

Fuelwood cutting must have substantially affected the Tombstone woodshed in the early 1880s. The federal government warned land claimants in the upper San Pedro Valley that it was illegal to cut wood on unpatented land claims other than to clear land for cultivation or for their own fuel needs (Rodgers 1965:53). Such warnings and records of arrests of violators are frequently encountered in the *Tombstone Epitaph* and the *Arizona Daily Star*. For example, the *Arizona Daily Star* (Apr. 23, 1884) reported:

152 *Primary Historic Human Impacts*

> There is considerable feeling . . . over the arrest of woodcutters. . . . The reason why is, simply because the people who are most interested are not aware of the wholesale destruction which is being made of the timber on government lands. The mesa tracts of southern Pima and eastern Cochise counties are being literally stripped of trees, so that shelter of stock will soon be unknown in these sections. There will be much more feeling than now when the facts are all known, and the action of the government will be supported by every citizen who has the good of the stock interest at heart. There is little enough timber on the plains of Arizona without having that which we have destroyed.

In some cases, the prohibition of woodcutting on the public domain led to the importation of large supplies of fuelwood from Sonora. For example, the *Arizona Daily Star* (Feb. 17, 1885) reported:

> Ten carloads of wood, cut in Sonora, were shipped to the Grand Central mine [Tombstone] last Tuesday. This comes from the fact that the United States forbids the cutting of wood on the public domain, and the duty on wood crossing the line is so very small that it amounts to almost nothing.

The scarcity of fuelwood due to overcutting was noted in local newspapers and in the surveyors' field notes of the U.S. General Land Office.[6] According to those field notes, the Dragoons, the Mules, the eastern slopes of the Huachucas, and southeastern slopes of the Whetstone Mountains were heavily cut for fuelwood. For example, Book 882 of the field notes describes Township 18S, Range 22E, just southwest of the Dragoon Mountains, as follows: "The surface of the township is composed of high rolling mesa land covered for the most part with good growth of grass. All timber has been cut and taken away for wood." Furthermore, the plat maps and field notes of the townships in the vicinity of Tombstone indicate that wood roads ran everywhere. Walter Lamb (n.d.), a resident of Tombstone in the late nineteenth century, observed: "There were wood roads fanning out from Tombstone like the veins of a leaf, some were just tracks, others well worn."

To assess the impact of historic fuelwood cutting on the Tombstone woodshed, Bahre and Hutchinson (1985:184–186) measured the incidence of cutting on the trees along thirty-seven belt transects in the evergreen woodlands of the Tombstone woodshed. For the most part, their evergreen woodland sites were located on National Forest lands in the Dragoon, Huachuca, and Whetstone mountains at elevations ranging between 4,800 feet and 6,000 feet.

Their age-structure study of the sites proved fruitless because the oaks were too hard for increment boring. Furthermore, they were unable to date the scars of past cutting. They found that few oaks had been killed by cutting, especially if they had been pollarded, and that 43 percent of the trees in the evergreen woodlands showed signs of cutting. Moreover, cut oaks had nearly three times more stems than uncut oaks. They concluded that more oaks were cut than their data indicate, because many cut stems had rotted off and the cambium had grown over old wounds, obliterating any sign of cutting.

The distribution, number, and size of several tree species may have been affected by woodcutting. For example, the scarcity of willows and cottonwood in photographs of riparian areas taken in the 1880s and 1890s may be partly due to their being cut to use as retort fuel, to kiln bricks, or to start mesquite and oak fires in the fireboxes of steam engines (*Arizona Daily Star* Oct. 24, 1880). There also appear to be fewer large oaks and junipers today because of woodcutting. Juniper does not have the BTU content of oak and mesquite, but it is an excellent fuel. Juniper also was much sought for fenceposts, and large juniper were cut for construction and mine timber in the 1880s and 1890s (*Daily Tombstone* Nov. 22, 1886). Some observers have doubted the impact of historical woodcutting on juniper because of the paucity of old juniper stumps, but the dead stumps themselves were much sought for fuel (Bahre and Hutchinson 1985:185).

I find no evidence that heavy fuelwood cutting between 1880 and 1940 resulted in major changes in the areal extent of the woodlands of southeastern Arizona. The evergreen woodlands, in particular, appear to have remained stable, most likely due to the pollarding of the oaks by woodcutters (Phillips 1912:13, 15). Apparently only deciduous oaks were killed by cutting. This finding is in opposition to the widely held notion that woodcutters commonly destroyed the woodlands around old mines (Brand 1933:45, 138–139; Sauer 1956:63–64; Parker 1979; Dobyns 1981). This idea may have originated from the fact that early woodcutters eradicated large areas of pinyon-juniper woodland in Nevada and New Mexico because they did not pollard the junipers, and pinyon rarely tolerates cutting (Lanner 1977; Young and Budy 1979; Samuels and Betancourt 1982). Today the volume of fuelwood standing in some parts of the evergreen woodlands is probably greater than at any time since 1870 because of fire suppression, controlled grazing, and greater Forest Service control of fuelwood cutting.

Only Robert West (1949:45–46, 116) has measured the effects of historic fuelwood cutting on the oak woodlands around an old mining district in the American Southwest and northern Mexico. He mapped areal changes in the plant cover that resulted largely from colonial fuelwood cutting for the mines in the Parral mining district of southern

Chihuahua, Mexico. Even though wood is still a major source of heating and cooking fuel there, West concludes:

> In spite of the 350-year period of exploitation, the areal extent of the oak forest probably has not greatly changed, although in places, as around Santa Barbara and San Francisco del Oro, the lower edges have been pushed back and invaded by grass and juniper for a distance of 4 miles. In composition, however, the forest appears to have been altered, with an increase in scrub forms and in juniper.

West further observes that in the Parral region large oaks and junipers are now scarce because of cutting, and that Emory oak survivors are not seedlings but coppice growths that have survived repeated cuttings, browsing, and severe drought. Finally, he observes the possibility of vigorous vegetation regrowth in the district since cessation of heavy fuelwood cutting about 1900.

In the 1880s and 1890s the semidesert woodlands of southeastern Arizona apparently suffered the same kind of degradation that affects the woodlands of many modern Third World countries where wood is still the primary fuel. Although the evergreen woodlands of Sonora are protected, modern dependence on fuelwood there possibly reflects the differences in the conservation of evergreen woodlands spanning the international boundary. For example, my preliminary investigations of the evergreen woodlands in Sonora indicate that tree densities are lower there than in Arizona and that more trees in Sonora have scars from cutting (Bahre 1984). In 1960, fuelwood was used for cooking in 64 percent of Sonora's occupied dwellings (México 1963:541); in 1970, the percentage dropped to 39 (México 1971:429); and by 1980, it was 23 percent (México 1983:285). Using Mexican census data, I estimated that from 1960 to 1970 approximately 660,000 cords of fuelwood were harvested just for cooking fuel in northeastern Sonora (Bahre 1984:108).

Based on current Forest Service cordage and regrowth rate figures, the demand for cordwood from the semidesert woodlands of southeastern Arizona since 1870 has equaled or exceeded the amounts available. The many trees that show obvious signs of cutting testify to the historic impact of cutting in southeastern Arizona.

In summary, while large junipers and oaks fell to the woodcutter's ax and the structure of the woodlands was affected, fuelwood cutting had little impact on the modern distribution of woodlands in southeastern Arizona, and the oak woodlands have apparently recovered since the 1930s and 1940s. There is, nevertheless, room for more investigation of the impact of fuelwood cutting on regional fire history, erosion, woodland structure and composition, and wildlife ecology.

8 Exotic Plant Introductions

Since 1870 many alien plants have been introduced into southeastern Arizona as ornamentals or to reseed degraded rangelands and control erosion. Several have become naturalized and have displaced natives, especially in grassland and riparian habitats. For the most part, however, exotics have found favorable habitats in areas of disturbance. The construction of roads, highways, and railroads, and of gas, telephone, and power transmission lines, in particular, has assisted in the spread of exotics. Land clearing for these rights-of-way has furnished disturbed and, in some cases, highly specialized habitats for plant establishment, and has facilitated the diffusion of exotics.

The number of exotics reported in the flora of Arizona has nearly doubled in almost fifty years. According to Kearney and Peebles (1942), approximately 190 of Arizona's 3,200 plant species in 1942 were introduced; today the number of adventives is 330 (Burgess et al. in press). In 1909, three exotic species were recorded for the flora of Tumamoc Hill on the outskirts of Tucson: filaree or afileria (*Erodium cicutarium*), barley (*Hordeum vulgare*), and Bermuda grass (*Cynodon dactylon*); today even in its largely undisturbed state (Tumamoc Hill has not been grazed by domestic livestock since 1907), 50 alien species (approximately 15 percent of the total flora of 346 species) are reported (Bowers and Turner 1985; Turner and Bowers 1987; Burgess et al. in press).

Some naturalized exotics, such as tamarisk or saltcedar (*Tamarix chinensis*) and Lehmann lovegrass (*Eragrostis lehmanniana*), have drastically altered composition of the grasslands and of riparian woodlands,[1] while others, such as London rocket (*Sisymbrium irio*), *Hordeum vulgare*, *Schismus arabicus*, *Schismus barbatus*, *Erodium cicutarium*, and red brome (*Bromus rubens*), are thoroughly integrated into some desertscrub and grassland communities. Certain exotic annuals are as well adapted to local desert conditions as are the natives. For example, seeds of *Bromus rubens* will germinate on as little as half an inch of rain, whereas most native winter annuals require at least an inch of rain for germination (Beatley 1966, 1967).

Some exotics, such as Russian thistle (*Salsola iberica*), *Schismus* spp., *Sisymbrium irio*, *Bromus rubens*, and *Erodium cicutarium*, are major colonizers of thousands of acres of recently abandoned agricultural land in southeastern Arizona (Karpiscak and Grosz 1979; Karpiscak 1980). Except for tamarisk, filaree, and Lehmann lovegrass, little research has been done on the distribution, ecology, and effects on the native vegetation of the more than 200 exotic species in southeastern Arizona. In some cases, exotic annuals are crowding out native perennial grasses; once the exotics are established, it is difficult for the natives to reestablish themselves. This is certainly the case in the California grasslands, where exotics now dominate (Biswell 1956; Naveh 1967; Frenkel 1970; Heady 1977). Furthermore, the invasion of desertscrub by exotic annuals may increase fire frequencies in that vegetation type (Rogers and Steele 1980; Rogers 1986). Even though the long-term impact of most exotics is unknown, several state and federal agencies continue to introduce exotic plants for erosion control and range improvement in southeastern Arizona. Table 8.1 lists the major exotics released for reseeding by the Tucson Plant Materials Center of the Soil Conservation Service in 1987.

Cattle ranchers and range managers have reseeded nearly 300 species of exotic and native plants in southeastern Arizona since the 1880s (Cox et al. 1984). For example, Bermuda grass was planted for erosion control along Ciénega Creek in 1902 (Potter 1902:2), while *Erodium* spp. (filaree), which was supposedly introduced into Arizona from California in the 1860s or 1870s, was planted as forage in the 1880s (*Arizona Daily Star* Feb. 10, 1880, June 13, 1886, Dec. 14, 1887).[2] A reporter for the *Arizona Daily Star* (Feb. 10, 1880) noted:

> ... there is a considerable belt of country on the eastern slope of the Santa Catarina [Catalina] Mountains that has got set [sic] with alfilare and is now green and blooming. This peculiar plant affords the most nutritious of feed for stock. The seed was introduced there a few years ago by a band of sheep from California having brought it in their wool and scattered it along the trails to their camps. We may expect to see it in a few years scattered all over our broad plains and we will find it a most valuable element in keeping stock, it growing even on ground where the grama will not grow.

Another report in the *Arizona Daily Star* (June 13, 1886) mentioned that ranchers seeded filaree: "The Leitch brothers are sowing alferia [sic] seed on their stock range. It would be well if some of the Pima County stockmen would follow their example." This exotic forage

Table 8.1. *Exotic Plant Varieties Recommended for Release by the Tucson Plant Materials Center of the Soil Conservation Service in 1987*

Name	Origin
"A-68" Lehmann lovegrass (*Eragrostis lehmanniana*)	South Africa
"A-84" and "Catalina" Boer lovegrass (*Eragrostis curvula* var. *conferta*)	South Africa
"Cochise" Atherstone lovegrass (*Eragrostis lehmanniana* × *E. trichophora*)	South Africa
"Palar" Wilman lovegrass (*Eragrostis superba*)	South Africa
"A67" weeping lovegrass (*Eragrostis curvula*)	Africa
"Corto" Australian saltbush (*Atriplex semibacata*)	Australia
"A-130" blue panicgrass (*Panicum antidotale*)	Australia
"Kuivato" and "Puhuima" Lehmann lovegrass (*Eragrostis lehmanniana*)	South Africa
Buffelgrass (*Cenchrus ciliaris*)	India or Africa
Yellow bluestem (*Bothriochloa ischaemum*)	Central Asia (Turkestan)
Mediterranean ricegrass (*Oryzopsis coerulescens*)	Morocco
Barley (*Hordeum vulgare*)	Abyssinia or Tibet
Mallees, Yates, boxes, and gums (*Eucalyptus* spp.)	Australia
Acacia (*Acacia notabilis*)	Australia
Thatchgrass (*Hyparrhenia hirta*)	Africa

plant was so significant to early ranchers in southeastern Arizona that in 1906, J. J. Thornber devoted an Arizona Agricultural Experiment Station Bulletin to the benefits of planting filaree on the range in Arizona. Noting that Hooker first planted *Erodium* in the Sulphur Springs-Aravaipa valleys in the 1860s, he points out that *Erodium cicutarium* yields up to two tons of hay per acre and makes first-rate forage (Thornber 1906).

The two exotics that have received the most interest from researchers of vegetation change in southeastern Arizona are tamarisk and Lehmann lovegrass. Tamarisk, a phreatophyte, has spread quickly on the floodplains of southeastern Arizona's major streams and presently dominates thousands of acres of former mesquite *bosques* and native riparian forests (Horton et al. 1960; Horton 1964, 1977; Robinson 1965; Harris 1966; Haase 1972; Turner 1974; Warren and Turner 1975; Ohmart and Anderson 1982; Minckley and Brown 1982; Rea 1983; Minckley and Clark 1984). For example, Haase (1972) estimated that about 50 percent of the lower Gila River bottomland was occupied by tamarisk in 1972. First introduced into the United States in the early 1800s for ornamental use (Horton 1964), tamarisk was naturalized by the 1930s along most of the rivers of the American Southwest. It attracted little attention, however, until it was realized that it created flood hazards by impeding stream flow and caused local water shortages by transpiring large amounts of water (Burkham 1970, 1972; Horton 1977). Its aggressive spread resulted in government eradication programs. One such program, the Gila River Phreatophyte Project, was established to study the amount of water lost through consumptive use by phreatophytes—mostly tamarisk and mesquite—and to measure the amount of water that might be salvaged by their removal (Burkham 1970, 1972; Turner 1974; Culler et al. 1982). However, because stands of tamarisk and mesquite are valuable as wildlife habitat, they may be safe from purposeful eradication. Nonetheless, declining groundwater levels have already taken a heavy toll on both species.

Tamarisk is well adapted to rapid colonization of areas of fresh alluvial deposition that have little or no plant cover (Horton et al. 1960; Warren and Turner 1975), conditions that, according to Harris (1966), were created by dam building and watershed disturbances in the Southwest. In addition, tamarisk germinates more readily in wet soils than does mesquite (Horton et al. 1960; Warren and Turner 1975). The areal extent of tamarisk in southeastern Arizona is likely to decrease, however, because of eradication programs, declining groundwater tables, and increases in water and soil salinity.

Lehmann lovegrass has widely invaded the grasslands of southeastern Arizona and by 1986 had come to dominate about 350,000 acres of grassland and former grassland in the San Pedro, Sulphur Springs, and Santa Cruz valleys (see Figure 8.1) (Freeman 1979; Cox and Ruyle 1986). Brought to Arizona in 1932, this South African, warm-season, perennial bunchgrass has been used extensively by the Soil Conservation Service and the Arizona State Department of Transportation to reseed degraded rangelands and transportation rights-of-way (Crider 1945; Williams 1964; Cox and Ruyle 1986). In 1940 lovegrass began to appear in areas that had not been reseeded (Cable 1971; Cox et al. 1987). According to Cox and Ruyle (1986), the spread of Lehmann lovegrass in southeastern Arizona is due largely to disturbed soil conditions and its ability to spread rapidly along transportation rights-of-way, which not only offer continuous corridors into the rangelands but also traverse environmental gradients. In addition, lovegrass has gained a competitive advantage because when cattle eat the native grasses, which they prefer to Leh-

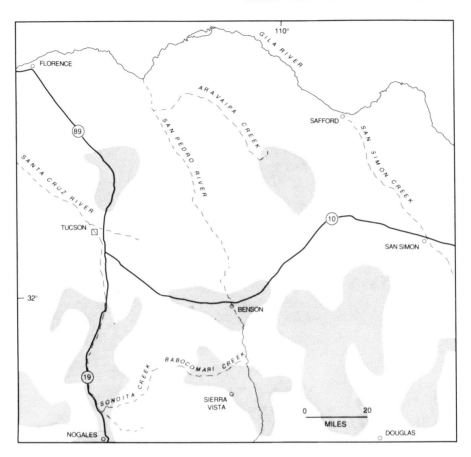

FIGURE 8.1. Major areas of Lehmann lovegrass in southeastern Arizona. *Source*: Cox and Ruyle 1986.

mann lovegrass during the summer growing season, native grass vigor is reduced (Humphrey 1959; Martin 1983).

The impact of Lehmann lovegrass on native plant and animal communities has been dramatic and largely negative. Bock et al. (1986) found that lovegrass communities on the Research Ranch near Elgin reduced native grass cover by 60 percent; lovegrass affected ten common plant species negatively and affected no species positively; grasshopper numbers were 40 percent less in lovegrass than in native grasses; and birds were more abundant in native growth. They concluded that sites dominated by native perennial grasses support a greater collected variety and abundance of indigenous plants and animals than do areas planted with or invaded by lovegrass, and that seeding exotic species is less desirable than giving indigenous species the time and opportunity to recover. Some researchers, however, believe that the spread of Lehmann lovegrass has peaked and that climatic and edaphic conditions will curtail it (Cox 1984a; Frasier et al. 1984; Cox and Ruyle 1986). Humphrey (1959:28) asks: "Mesquite has made a good start on many of our southern Arizona ranges in driving out our native grasses; is Lehmann lovegrass going to finish the job?"

Other exotics favored for forage and fairly widespread are *Bromus rubens, Chloris virgata, Eragrostis chloromelas, Erodium cicutarium, Eragrostis curvula, Panicum antidotale,* and *Sorghum halapense.*

The exotics that have taken over thousands of acres of abandoned agricultural lands in southeastern Arizona have interested researchers, especially from the Office of Arid Lands Studies at the University of Arizona. They are studying the economic and environmental problems resulting from the abandonment of agricultural land as well as the feasibility of harvesting certain weedy exotics such as Russian thistle or tumbleweed, which has been harvested to make fuel logs.

Another exotic expanding its range in southern Arizona at elevations below 2,000 feet is buffelgrass (*Cenchrus ciliaris*). According to Tony Burgess (interview, Sept. 1988), a new strain of buffelgrass (introduced by the Soil Conservation Service) could invade much of southeastern Arizona below 5,000 feet in elevation.

Even though exotic plants have become a significant part of the modern landscape of southeastern Arizona, little is known about their distribution, ecology, or impact on native plants. Almost all of the exotics in Arizona have been introduced since the turn of the century, and their spread is due mostly to human interference. In spite of the fact that some exotic plants have negatively affected the ecology of native plant and animal species, governmental agencies continue to seed exotics for range improvement and soil stabilization.

9 Agriculture, Logging, and Haying

AGRICULTURE AND GROUNDWATER PUMPING

In many ways, agriculture has affected the modern vegetational landscape of southeastern Arizona. For example, agricultural clearing has resulted in the extirpation of thousands of acres of desertscrub, grassland, and riparian wetland; created favorable habitats for weeds; caused erosion; and, along with irrigation works and other types of bottomland disturbance, resulted in or exacerbated flooding and stream entrenching. In addition, the heavy overdraft of groundwater for irrigation (in 1986, according to a U.S. Department of Agriculture and Arizona Water Commission report, 87 percent of all water depletion in Arizona was due to irrigation) has led to rapid lowering of groundwater depths and, consequently, to the cessation of surface flow in certain streams and to the destruction of thousands of acres of mesquite, tamarisk, and other phreatophytes. Changes in soil and water salinity and alkalinity due to irrigation have also affected local native vegetation.

About 3 percent of southeastern Arizona, mostly floodplain and valley bottomland, has been cleared for farming. The area of cropped and abandoned farmland far exceeds the combined areas of ponderosa pine and mixed-conifer forest and riparian wetland. In addition to the areas cleared for cropland are lands cleared for highways, roads, ditches, and settlements associated with agricultural development. These account for about 7 percent of the 3 percent cleared for cropping.

Before development of large-scale groundwater irrigation in the 1940s, agriculture in southeastern Arizona depended primarily on surface water for irrigation on the floodplains of the major perennial rivers and creeks: the Santa Cruz, San Pedro, Sonoita, Babocomari, San Simon, Rillito, and Aravaipa.

In the early 1870s, irrigated agriculture was the primary focus of most settlement in southeastern Arizona and was concentrated on the

Gila River near the Pima villages; at San Jose near the mouth of San Simon Creek; on the Santa Cruz River between Calabasas and Tucson; on Sonoita Creek from its mouth to Crittenden; on Babocomari Creek near Camp Wallen; on the San Pedro River at San Pedro, Tres Alamos, and Redington; and along parts of Rillito and Aravaipa creeks (Hinton 1878; *Arizona Daily Star* Aug. 31, 1879; U.S. Congress, House 1879; Meyers 1911; Schwennesen 1919; Barnes 1935; Rodgers 1965). By 1870 all of the following settlements were either adjacent to areas of irrigated agriculture or were themselves agricultural villages: Calabasas, Crittenden, Rillito, San Pedro, Florence, San Xavier, Saginaw, Adamsville, Camp Grant, Tubac, and Tucson (U.S. Bureau of the Census 1872) (see Figure 9.1).

Mormon farmers established St. David in 1877 and Hereford in 1879 in the upper San Pedro Valley; and Pima in 1879, Thatcher in 1880, and Bryce in 1883 in the upper Gila Valley (McClintock 1921). In the late 1870s and early 1880s irrigated agriculture was expanded in the Florence area and along the San Pedro River from its mouth south to Mammoth (U.S. Congress, House 1881). In 1881 there was also some farming near San Simon in the San Simon Valley and at Pantano on the mouth of Ciénega Creek just east of Tucson (ibid.) (see Figure 9.1).

In 1882 artesian wells were discovered at St. David (*Arizona Weekly Star* Nov. 11, 1883), and soon after, artesian wells were developed near Thatcher, Safford, San Bernardino, San Simon, and Bowie (Skinner 1903; Forbes 1911; *Artesian Belt* 1914; Schwennesen 1919). Although the flow from these wells was insufficient to irrigate large areas, discovery of artesian wells, along with subjugation of the Apache, led to agricultural expansion in the Sulphur Springs and San Simon valleys.

Some dry or rain-fed farming was also practiced by homesteaders from the 1880s to the 1920s, especially in the Sulphur Springs Valley around Kansas Settlement and in the Sonoita-Elgin area. Dry farming was largely unsuccessful, however, and did not involve a large acreage in southeastern Arizona (Fraizer n.d.; Scott 1914; Rodgers 1965; Bahre 1977).

Initiation of major canal and dam projects to expand irrigation development, especially along the Gila River, paralleled the discovery of artesian wells. For example, thousands of acres were irrigated before 1900 in the Florence and Casa Grande areas from the Florence Canal (Florence-Casa Grande Canal) and the Casa Grande Reservoir (Picacho Reservoir). The *Arizona Weekly Enterprise* (Aug. 10, 1889) noted that "... over 6,000 acres of land were supplied by water by the canal [Flor-

FIGURE 9.1. Irrigated cropland along the Santa Cruz River in Tucson, 1882. Photograph by C. E. Watkins, courtesy of the Huntington Library.

ence Canal] and were cultivated for the first time." In the same year, the *Southwestern Stockman* (Sept. 14, 1889) reported that 300 acres were under cultivation in the San Simon Valley, along with 2,000 acres in the Sulphur Springs Valley and adjacent foothills, largely from artesian wells; windmills were also used to pump groundwater for irrigation.

Steam-powered engines were used in some locations to pump groundwater for irrigation in the 1890s, but because of high operation costs, they were employed only where groundwater was near the surface. Woodward (1904), who studied irrigation pumping costs in the vicinity of Tucson around 1900, observed that one cord of mesquite was required to run the steam engine on a shallow-draft water pump for ten hours; and of the nine pumps he tested in 1904, five relied solely on wood for fuel, three on crude oil, and one on electricity. Steam-powered pumps were used for irrigation along Rillito Creek up to 1910 (Smith 1910).

The acreage irrigated by surface water and well water in southeastern Arizona in 1910 was minuscule compared with the acreage irrigated by groundwater pumping in the 1970s. For example, the census of agriculture in 1910 (U.S. Bureau of the Census 1913:621) notes that of the 4,900 acres irrigated in Cochise County in 1909, 3,094 acres were irrigated from surface waters, while the remaining acreage was irrigated from artesian wells, steam pumps, and windmills. In 1911, 5,800 acres were irrigated along the San Pedro River from Benson to St. David and 6,000 acres were irrigated along the Santa Cruz River from Nogales to Tucson (Forbes 1911). In 1928 completion of Coolidge Dam and San Carlos Reservoir on the Gila River led to major expansion of irrigated agriculture on the middle Gila River floodplain.

The major boom in irrigated agriculture came during and shortly after World War II, when the development and acquisition of efficient, low-cost groundwater pumps contributed to the rapid expansion of irrigated agriculture in the Santa Cruz, San Pedro, Sulphur Springs, and San Simon valleys. In the last two valleys, thousands of acres of desertscrub and grassland were converted into some of the most productive cropland in the American Southwest. According to the U.S. Department of Agriculture and the Arizona Water Commission (1986), the annual volume of water pumped for all uses increased from 200,000 acre-feet in 1920 to 1.5 million acre-feet in 1940, and from 4.5 million acre-feet in 1953 to 5 million acre-feet in 1986.

Although the number of acres irrigated from surface waters has remained fairly constant since the mid-1970s, the total acreage in southeastern Arizona irrigated by groundwater has declined precipitously since 1973, when higher fuel costs, greater depths to groundwater, unfavorable economic conditions for farming, and the higher value of urban land and water uses sent the agricultural industry dependent on groundwater into decline (Frederick 1982; Meitl et al. 1983; Towne 1986; Kennedy et al. 1986). Meitl et al. (1983) estimate that about 300,000 acres of agricultural land were retired in southern Arizona between 1973 and 1983. According to the Arizona Crop and Livestock Reporting Service (1985), the total planted area in Arizona declined by 66 percent from 1977 to 1983. Cochise County experienced the greatest decline in irrigated farmland of any county in Arizona. Its irrigated agriculture had increased rapidly from about 9,800 acres in 1948 to 134,000 acres in the mid-1970s, then declined to less than 41,500 acres in 1984 (Arizona Crop and Livestock Reporting Service 1985; Towne 1986). By the year 2020 the Arizona Water Commission anticipates the virtual disappearance of irrigated agriculture in Cochise and Pima counties (Meitl et al. 1983). As an indication of this trend, the city of

Tucson is purchasing adjacent farmland, especially in the Avra Valley, and is retiring it from agriculture for the water rights; likewise the city of Scottsdale is purchasing farmland in the Coolidge area for water rights.

The abandonment of thousands of acres of formerly irrigated cropland has had serious economic and ecological consequences. With little vegetative cover to retain topsoil, dust blows; weeds, such as Russian thistle and carelessweed (*Amaranthus palmeri*), encroach; and grazing potential is lost (Karpiscak 1980). More emphasis is being placed on reseeding abandoned fields with alien grasses or drought-resistant crops.

Olmstead (1919), Bryan (1925), Cooke and Reeves (1976), Dobyns (1981), Rea (1983), and Waters (1988) are among many who suggest that agricultural clearing and irrigation works have exacerbated, if not caused, flooding in the region.[1] For example, Olmstead (1919) blames entrenching along the San Simon River on the digging of the Solomonville drainage ditch, and Dobyns (1981) gives numerous examples of floods exacerbated by irrigation works along the Santa Cruz and upper Gila rivers. Harris (1966) and others attribute the spread of tamarisk along the major rivers of the American Southwest, particularly the Gila River and its tributaries, to human disturbance of the bottomland, especially to stabilization of stream flows by dam construction (Coolidge Dam on the Gila River was completed in 1930).

Declining groundwater levels are frequently noted to have had a major impact on floodplain, riparian, and emergent vegetation in southeastern Arizona (Bryan 1928; Brandt 1951; McQueen and Miller 1972; Dobyns 1981; Brown 1982; Ohmart and Anderson 1982; Rea 1983; Hendrickson and Minckley 1984). Yet there is no study that assesses the extent and impact of declining groundwater levels on regional vegetation cover or stream flow.

Although use of groundwater for crop production has declined rapidly, demand for it for domestic and other purposes has increased greatly (Bowden 1977). Between 1965 and 1970, water levels in much of the Santa Cruz Valley dropped by as much as 50 feet, and since 1947 declines of more than 125 feet have occurred (Matlock and Davis 1972). Groundwater levels throughout the region are still falling.

Because cottonwoods and willows are shallow-rooted phreatophytes, they are the first trees to be killed by declining water tables. Mesquite and tamarisk, which have roots that can reach forty feet or more, are able to survive longer (Parker and Martin 1952; Phillips 1963; Minckley and Clark 1984). Mesquite does not grow in waterlogged soils, however, and it probably utilizes soil moisture above the water table far more than water from the saturated zone (McQueen and Miller 1972). On the other hand, tamarisk requires long periods in saturated soil to germinate and establish itself (Warren and Turner 1975).

Thousands of large, old mesquite have perished since the 1940s because of declining groundwater levels in mesquite *bosques*, such as those at San Xavier del Bac, Casa Grande ruins, New York Thicket (Komatke), and New York Hill. Judd et al., who studied the die-off of mesquite at Casa Grande National Monument in the 1960s, conclude (1971:159):

> The cause of the lethal decline of mesquite trees on the monument appears to involve a number of factors, each contributing to the end result. Decline of the water table and mistletoe infestation may be the major contributors with age of trees, insect infestation, and natural successional process as secondary factors.

No doubt many acres of riparian forest and mesquite *bosques* have perished as a result of groundwater overdrafts; even the expansion of tamarisk has been curtailed by falling groundwater levels. Nevertheless, vegetation changes due to declining groundwater levels in southeastern Arizona remain mostly unexplained.

LOGGING

Although nearly all of the ponderosa pine and mixed-conifer forests in the study area have been logged to some extent at one time or another, historic records indicate significant logging occurred only in the Chiricahua, Pinaleño, Huachuca, Santa Rita, and Santa Catalina mountains—and then mostly before 1900 (see Figure 9.2). Since that time limited logging has continued in the Santa Catalina and Pinaleño mountains.

Of the many researchers who have studied the ecology and forest history of the scarce ponderosa pine and mixed-conifer forests of southeastern Arizona, only Kellogg (1902a, 1902b, 1902c), Hoffmeister and Goodpaster (1954), Wallmo (1955), Harrison (1972), Matheny (1975), and Weech (1979) have noted the impact of logging, usually in fleeting comments on nineteenth-century logging in the Catalina, Santa Rita, Huachuca, Chiricahua, and Pinaleño mountains. For example, Wallmo, in his study of the vegetation of the Huachuca Mountains, writes (1955:479):

FIGURE 9.2. Ross Sawmill in Morse Canyon in the Chiricahua Mountains circa 1885. Photograph by C. S. Fly, courtesy of the Bisbee Mining and Historical Museum, Brophy Collection.

In the early days considerable logging was carried on in the conifer forests of the Huachucas, mainly on the "Reef," a less precipitous area above the quartzite cliffs and below Carr Peak. Extensive logging was also done below Ramsey Peak on the military reservation. A devastating fire near the turn of the century terminated logging in the Reef area and no commercial operations have since been conducted. Mexican white pine, Apache pine, and Douglas fir were the major timber species.

Harrison's study, "The Santa Catalinas—A Description and History" (1972), is the most comprehensive history of logging for any range in southeastern Arizona, but it does not discuss the areal extent or ecological impact of logging in that range.

According to the historic record, the first mountain range in southeastern Arizona extensively logged by Anglos before 1870 was the Santa Ritas, where pine and Douglas fir were cut (largely in Sawmill and

Madera canyons) in the 1850s, 1860s, and 1870s to supply mine timbers and lumber to Tubac, Tucson, and nearby settlements (Matheny 1975). Before the 1850s Amerinds, Spaniards, and Mexicans undoubtedly cut timber in several high-elevation conifer forests, but the amounts harvested were probably insignificant.

There are scores of references to sawmills and logging operations in the nineteenth-century newspapers of Tucson, Florence, Tombstone, and Bisbee. These reports shed little light on the amounts of sawtimber removed or on the ecological impact of lumbering on the forests. They do indicate, however, that logging operations were larger and more widespread than commonly acknowledged. Below are typical newspaper accounts of logging in the most extensively harvested ranges.

Huachuca Mountains

Three saw mills are in full blast on the Huachucas. . . . (*Arizona Weekly Citizen* Oct. 25, 1879)

Mr. F. M. Tanner of Hayes and Tanner, owners of the steam saw mill on the west side of the Huachuca Mountains, is in Tucson and favored us with a call last evening. Their old mill, it will be remembered, was some time since almost entirely destroyed by an explosion of the boiler. The machinery, engine, saws, etc., have all been replaced by new material, and the new mill has been running for about a week now. It has a capacity of 10,000 feet per day, and it cuts steadily 7,000 to 8,000 feet, for all of which a ready market is found in the Harshaw and Washington districts.

The demand for lumber has been greater than they have been able to supply, but since their new mill is in operation all orders can be filled promptly. (*Tucson Daily Record* May 13, 1880)

January 1879 . . . mill started up at Saw Mill Canyon [east side of Huachucas] after exhausting timber in that vicinity, moved to the top of the mountains, 8,000 feet altitude. . . . This mill for a time supplied all the lumber that was to be had in the country. The mill had a capacity of 10,000 feet per day, and was kept constantly running without intermission. . . . This lumber is used for ore houses, mill and hoisting work buildings, boarding houses, mining timbers, and other purposes. (*Weekly Nugget* June 10, 1880)

Turner and Campbell's Saw Mill . . . one mile above Turnerville [east side of the Huachucas in Ramsey Canyon] commenced cutting lumber on the 10th of May and has turned out since that time over 90,000 feet. . . . upwards of 50 men employed . . . it is estimated that the lumber capacity of the canyon will be 4,000,000 feet. . . . the Huachuca Saw Mill is situated high up on the northern side of McCloskey Canyon. It was built in December 1878. . . . a 24 hp engine is the motor which drives a 60-inch saw and the capacity of the mill is from 8,000 to 10,000 feet a day. To meet the urgent demands of customers, a night shift is to be put on soon and the output will be nearly, if not quite, doubled. Since commencing work, 1,750,000 feet of lumber have been turned on the market, and it is estimated . . . 3,500,000 feet more are available. (*Tombstone Weekly Epitaph* June 12, 1880)

Pinaleño (Graham) Mountains

. . . the new road leads to a dense growth of timber composed of white and sugar pine, red and white fir; sufficient not only to supply the wants of this post [Camp Grant] for all time to come,' it may be said to be inexhaustible; . . . trees of any and all sizes can be obtained, some measuring as much as five feet through and from fifty to sixty feet in length clear of knots. The timbered part of the summit [Mt. Graham] is from one to three miles in length. . . . (*Arizona Weekly Citizen* Nov. 8, 1873)

Frye's Sawmill, located in the Graham Mountains about 20 miles from Solomonville, has changed hands, Joseph Alfred having purchasing it from Mr. Frye. (*Arizona Daily Star* Feb. 12, 1890)

Santa Catalina Mountains

We noticed some excellent saw logs brought from the San [sic] Catalina pinery by Mr. Leon and others. James Lee also has the first installment of 20,000 feet he is hauling in. . . . (*Arizona Weekly Citizen* Sept. 20, 1873)

On Monday about 2 o'clock p.m. thirty Apaches attacked five Mexicans who were engaged cutting timber in the Santa Catarina [Catalina] Mountains, about thirty miles northeast of Tucson. . . . (*Arizona Weekly Citizen* June 20, 1874)

. . . the Santa Catarina [Catalina] Copper Company. . . . the saw mill accompanying the plant will furnish all necessary lumber. The supply of timber in the mountains above is said to be one of the finest belts in Arizona and practically inexhaustible. The lumber for present building purposes must be taken from Tucson. . . . The charcoal will also be burned on the ground, 1,000,000 bushels being already contracted for. . . . (*Arizona Weekly Enterprise* Feb. 18, 1882)

. . . the saw mill [in the Santa Catalinas] turns out 5,000 feet of lumber per day and since it commenced work has paid its cost three times over. The timber field is one and a half miles from the works with a good road that cost $2500. The area of pine timber reached by this road is thirty-six square miles. There is a probability that Tucson will be largely supplied from the Santa Catalina mill, especially for mills, hoisting works, and mines as the facilities for furnishing this class of lumber is [sic] better than from California, as well as cheaper. . . . (*Arizona Daily Star* May 16, 1882)

Mr. Clifton, of Mammoth, has contracted to furnish timbers for the [Mammoth] Mine. He receives $25 per 1000 for round and $35 per 1000 for square timbers. He will haul the timbers from the Santa Catalinas. . . . (*Arizona Weekly Enterprise* Aug. 7, 1886)

Santa Rita Mountains

The saw mill of Preston and Heslip in the Santa Rita Mountains is running constantly sawing logs cut in Saw Mill Cañon. They have just let a contract to Herbert Brown to deliver a large number of logs, monthly; which now shows the pressing demand for lumber at the various mines. Santa Rita Mountains produce quantities of timber of the finest quality. (*Arizona Weekly Star* June 5, 1879)

Chiricahua Mountains

Chiricahua Saw mill—this mill, property of Phillip Morse and Co., has now been running nine months, and during that time over 1,000,000 feet of lumber has been shipped, most of which has been consumed in Tombstone and adjacent mills. The mill is now turning out 50,000 feet per week. (*Arizona Weekly Star* May 27, 1880)

Jacob Hoke—Dealer in lumber, keeps supplies from Morse Mill in Chiricahua Mountains and deals in siding, tongue and grooved flooring, planed lumber, etc. (*Weekly Nugget* June 10, 1880)

Attention Ranchmen! shakes, shingles, etc. for sale from mill at head of Morse's Canyon. Juniper timber 30 feet in length. (*Daily Tombstone* Nov. 22, 1886)

The Ross Mill in the Chiricahua Mountains is turning out in the neighborhood of 20,000 feet of lumber per day.... (*Arizona Daily Star* Oct. 24, 1889)

After the Southern Pacific arrived in Arizona in 1881, the bulk of the lumber used was imported from California, Oregon, and Washington. Nevertheless, many miners preferred local wood for mine timbers in wet ground (*Tombstone Prospector* Oct. 19, 1889; *Arizona Weekly Enterprise* Apr. 5, 1890) and until the 1890s, some mines still used local sawtimber, particularly from the Catalina, Pinaleño, and Chiricahua mountains. The Copper Queen at Bisbee depended on sawtimber from the Chiricahuas in its mining operations until the 1890s (*Bisbee Review* Aug. 8, 1923). The Ross sawmill in the Chiricahuas, from which the Copper Queen purchased its timbers, was frequently cited for illegally cutting on public lands. For example, the *Arizona Weekly Star* (Nov. 12, 1885) reported:

Timber Depredations ... Parties from the Chiricahua Mountains say there is a large amount of timber being cut on government land in the mountains about fifty miles from Bisbee. If this is so, this is work for the U.S. timber agents. It is said the pine timber is being literally cleared off the government land, and if the depredations continue at the present rate, there will not be a stick left on that side of the mountain.

Eventually the government filed both criminal and civil actions against Ross and the Copper Queen (*Bisbee Review* Aug. 8, 1923). "The cutting in the famous 'Copper Queen' case, which was decided recently in favor of the company, took place in the Chiricahua Mountains 7 to 12 years ago, and the cut-over area is completely skinned" (Kellogg 1902a:505).

The impact of clear-cutting by early logging operations, especially in the Chiricahua, Pinaleño, and Huachuca mountains, is unknown. However, we can assume that clear-cutting would have led to changes in the structure of the forests, with ensuing young growth stands and possibly less diversity in both animal and plant species. Clear-cutting may also have increased decomposition, nitrification, runoff, and erosion. Historic records indicate that, among other places, major logging operations were carried out in Barfoot Park, Rock Canyon, Pinery Canyon, Downing Canyon, Morse Canyon, and Rucker Canyon in the Chiricahuas; Sawmill Canyon, Ramsey Canyon, Garden (Tanner's) Canyon, Carr Canyon, and Sunnyside Canyon in the Huachucas; between Mt. Lemmon and Green Mountain across the Palisades in the Catalinas; the two Sawmills, Madera, and Gardner canyons in the Santa Ritas; and in Nuttall, Shingle Mill, Ash, Frye, Jacobson, Marijilda, Tripp, and Webb canyons, and all around Mt. Graham between Grandview and Heliograph peaks, in the Pinaleños.

According to Coronado National Forest records, 455 thousand board feet (MBF) of sawtimber were removed from the study area in 1987—an insignificant amount, especially compared with nineteenth-century harvests or with the 7,208 MBF removed in 1980.[2] All of the sawtimber harvested since 1971 has come from the Pinaleño and Santa Catalina mountains (John Turner, National Forest Service, letter to the author, Nov. 1987). Logging waned in the other mountain ranges after 1900, and no major harvests have been taken from the Chiricahua or Santa Rita mountains for several decades (John Turner, National Forest Service, interview, Mar. 1982). Kellogg (1902a:504) writes: "... all the good timber was cut out [of the Huachuca Mountains] years ago." That the Pinaleños had been heavily logged before the turn of the century is attested to in a report by Kellogg (1902c:10):

> A great deal of cutting has been done [referring to the Grahams or Pinaleños]. In fact, it may be said that at one time or another, saw mills have been located in every accessible canyon, and at present there is very little good timber below 7,500 feet that can be reached easily. One-hundred thousand feet of lumber at Jacobsen's mill has been confiscated by the General Land Office, and will be sold at auction in a few days. These mills of course, like many others, have operated illegally for years; since they have been shut down the local price of timber is $40 per 1000 where it was formerly $25.

Approximately 23 million board feet (MMBF) were removed from the Pinaleño Mountains between 1890 and 1946 and 40 MMBF between 1962 and 1972 (John Turner, letter, Nov. 1987). The harvest between 1947 and 1961 was not tabulated. The total harvest of 63 MMBF between 1890–1946 and 1962–1972 in the Pinaleños would amount to clear-cutting about 10,161 acres or 15.9 square miles of ponderosa pine forest in the range under present conditions, a little more than one-third of the present area of ponderosa pine and mixed-conifer forest.[3] This figure is based on Alexander's (1974:26) estimate that ponderosa pine stands in Arizona average about 6.2 MBF per acre. Some fully stocked stands on good sites, however, are capable of yields of 25 to 35 MBF per acre (ibid.) Yields from virgin stands of big trees when these forests were first logged, however, would have been much greater; therefore the total area cut to reach a total harvest of 63 MMBF was probably much smaller than 10,000 acres. The average of 6.2 MBF per acre, however, is 1.2 MBF greater than the figure of 5 MBF estimated for the Chiricahua Mountains by Kellogg (1902a:503) in 1902. If the harvest between 1947 and 1961 and after 1972 were included, the total amount harvested in the Pinaleño would be much greater than 63 MMBF. Unfortunately, I was unable to obtain data on the total harvest from the Santa Catalina Mountains or other ranges.

During the 1980s the annual harvest for the Santa Catalina Mountains was 230 MBF and for the Pinaleños 225 MBF. These figures are not likely to increase because local demand is limited and forest resources are being preserved for recreation and wildlife (John Turner, letter, Nov. 1987). The harvest is largely relegated to salvage logging, removing diseased and insect-infested trees, or clearing for roads, camping areas, and skiing courses.

In spite of the significant amount of sawtimber that has been harvested from the ponderosa pine and mixed-conifer forests of southeastern Arizona, there are no studies that estimate the total amounts of sawtimber harvested or the areal extent and impact of historic logging.

WILD HAY HARVESTING

The harvesting of wild or mesa hay in the plains grasslands and desert grasslands of the major valleys was once a significant economic activity in southeastern Arizona (Bahre 1987) (see Figure 9.3). Between 1850 and 1920, local wild hay met much of the demand for horse, mule, and burro fodder in cities, towns, mining camps, and military posts (*Ari-*

FIGURE 9.3. Stacks of wild hay on the Santa Rita Experimental Range, 1902. Courtesy of the U.S. Forest Service and S. Clark Martin.

zona Weekly Star May 23, 1878; Martineau 1885; Griffiths 1901; Smith 1910; Wooton 1916:23–28; Rockfellow 1955:60; Willson 1966). In addition, ranchers put up large amounts of hay for livestock feed. So significant was wild hay harvesting in the past that hay roads and hay corrals were frequently noted in the surveyors' field notes, especially in Cochise County (Surveyors' Field Notes, books 942 [1898], 943 [1898], and 958 [1885]). Today most of the valleys in which wild hay was cut have been cleared for irrigated agriculture and urban development, or have been overgrazed and taken over by brush and scrubby trees—so much so, that at present it is difficult to believe that one contractor in 1902 had twenty mowing machines cutting hay near Fort Grant in the Sulphur Springs Valley (*Arizona Daily Star* Oct. 14, 1902).

For the most part, wild hay was mowed in flat, unfenced lands, usually from October to May during years of above-normal summer precipitation when grasses were luxuriant (*Arizona Weekly Citizen* May 2, 1874; *Tombstone Epitaph* Sept. 17, 1881; *Tombstone Daily Nugget* May 9, 1882; *Arizona Daily Star* Oct. 12, 1889, Aug. 18, 1904; Gray 1940:119). In addition, Mexicans and Indians (mostly Apaches and Tohono O'odham) used sickles, knives, and hoes to cut wild hay (Rockfellow 1955:60; Hastings 1959:61; Willson 1966). The preferred hay grasses were grama, galleta, sacaton, fingergrass, threeawn, millet, vine mesquite, and little bluestem. The gramas, however, were the most favored (Toumey 1891a; Griffiths 1901:14–15).

The largest amounts of wild hay were probably cut before 1899, the year of the first census of wild hay harvesting in Arizona. According to the censuses, Arizona farmers harvested 9,524 tons of wild hay in 1899, 8,168 tons in 1909, and 7,802 tons in 1919. Furthermore, between 1909 and 1919, about 60 percent of this hay was harvested in southern Arizona, nearly 80 percent of it in Cochise County. However, these census figures do not include the amounts of hay harvested by ranchers, Indians, and entrepreneurs other than farmers. Yields per acre ranged from 0.86 to 0.92 ton, although yields up to 2 tons per acre are reported for riparian areas (*Pinal Drill* Sept. 25, 1880; U.S. Bureau of the Census 1902:218, 1913:74, 1922:273; Bahre 1987).

Demand for wild hay in southeastern Arizona declined rapidly after 1910, mostly because of the development of motorized transport, growth of irrigated agriculture, overgrazing of grasslands, fencing of the public domain, expansion of private landholdings, clearing of native vegetation in valleys, uncertainty of the wild hay crop, and invasion of the grasslands by woody plants. The increase of woody plants, in particular, impeded the mowing of grass in many formerly harvested areas. For example, Government Draw, near Tombstone, which at one time supplied tons of wild hay for Bisbee and Tombstone (*Arizona Daily Star* Sept. 8, 1905), is now mostly covered by mesquite and acacia. Likewise, once-extensive grasslands along the lower slopes of the Santa Rita, Dragoon, Huachuca, and Mule mountains are now covered by woody plants (Robert 1869) (see Figure 6.3).

Wild hay harvesting was one of many historic land uses that either point to or may have resulted in major vegetation changes in southeastern Arizona's semidesert lands. It was abandoned in part because wide expanses of grassland for mowing—especially in the major valleys—are now scarce. That wild hay harvesting itself led to major vegetation changes is conjectural; nevertheless, two newspaper accounts mention damage caused by hay cutting in the Fort Grant and Silver Bell areas (*Arizona Weekly Citizen* Nov. 8, 1873; *Arizona Daily Star* Feb. 23, 1882), and one researcher (Smith 1910:98) believes that hay harvesting contributed significantly to erosion along Rillito Creek in the 1880s. The newspaper account of the harm caused to the grasslands of the Silver Bell area by hoeing follows:

A party just in from Silver Bell District reports that the vast plain of gramma [sic] grass west of Tucson is being dug out by the roots, thus totally destroying the hope of the grass starting where it has been cut out. Many tons have been dug out by the roots and brought to this city for sale. The gramma [sic] grass of Arizona is the finest pasturage known, and is a source of great wealth in the growth of stock. This grass can be cut without killing the roots, and to this there cannot be urged any objection, and while our solons are passing game laws, it would be wise for them to look after our pasture lands. Gentlemen, give us a law to stop what might be termed barbarous practice of totally destroying our native pasture lands. Unless something is done, the gramma [sic] grass will soon be a thing of the past in Arizona. (*Arizona Daily Star* Feb. 23, 1882)

10 Patterns and Factors of Change

That vegetation has undergone several changes in southeastern Arizona since 1870 is evident from a study of the General Land Office surveyors' field notes, repeat ground and aerial photography, permanent plot studies, and historical diaries and reminiscences of the nineteenth century. The evidence also indicates that some cyclic and certain directional changes are confined to individual major vegetation types. This chapter summarizes the extent and causes of these changes in each major vegetation type in southeastern Arizona.

RIPARIAN WETLANDS

Since 1870 riparian wetlands have lost their once-luxuriant aspect. The rivers and streams that flowed in the nineteenth century now have a much decreased surface flow, deeply entrenched channels in parts, and banks that support little of the native vegetation that once dominated. The *ciénagas*, mesquite *bosques*, and forests of cottonwood and willow that once punctuated the floodplains and stream banks have been badly degraded or eradicated; consequently, they have either disappeared or been replaced by tamarisk (*Tamarix chinensis*) and other exotics.

Although no large mesquite *bosque* appears to have escaped some type of human impact, several large forests remain along the middle Gila and San Pedro rivers. But these protected *bosques* are also threatened by declining groundwater tables and have understories dominated by exotics. Furthermore, due to grazing and stabilized stream flow, many remaining stands of cottonwoods and sycamores are even-aged and are not regenerating (Glinski 1977; Minckley and Brown 1982:270–271).

A fascinating illustration of the change in riparian habitats comes from the prehistoric Indian site of Quiburi on the San Pedro River just west of the photo station in Figure 4.2. At that site, archaeologists removed from a 300-year-old trash pit the caudal vertebrae of a Colorado squawfish (*Ptychocheilus lucius*) nearly five feet long (DiPeso 1953:236; Miller 1961:375). Today the San Pedro River near Quiburi is intermittent and no natural water body within 100 miles could sustain such a large fish.

Along streams where the flow has not been stabilized by dams and major irrigation works, such as along sections of the Santa Cruz and San Pedro rivers, the riparian vegetation is in a constant state of flux or "perpetual succession" due to shifts in the stream channel and such periodic disturbances as flood scouring, inundation, and desiccation (Campbell and Green 1968; Everitt 1968; Applegate 1981). That riparian woodlands in these habitats are particularly dynamic is demonstrated by the rapid changes in the cottonwood-willow communities seen in repeat ground and aerial photographs of the San Pedro and Santa Cruz rivers (Hastings and Turner 1965; Bahre and Bradbury 1978; Gehlbach 1981; Applegate 1981) (see Figures 4.2, 4.3, and 4.12). Apparently these galeria forests are maintained by periodic spring floods (Campbell and Green 1968; Applegate 1981), as evidenced by the new cottonwood and willow communities that sprang up along portions of the San Pedro River following the spring floods of 1962 and 1967.

The modern extent of wetlands and perennial streamflow in southeastern Arizona has been mapped by Brown et al. (1981). They have used historic records to contrast present-day and nineteenth-century wetland distributions. Even though the number and extent of wetland habitats have decreased greatly since 1870, southeastern Arizona still has the greatest known past and present abundance of *ciénagas* in the American Southwest (Hendrickson and Minckley 1984). Although modern *ciénaga* habitats now persist largely only in headwater areas and around major springs (ibid.:131), they are much reduced in size, variously modified, and often artificially maintained. In fact, all of the *ciénagas* I visited in southeastern Arizona have been badly disturbed at one time or another and usually have only a semblance of "naturalness." As pointed out earlier, many of the present macrophytic plant taxa of *ciénaga* habitats in southeastern Arizona are exotic. This alone indicates that the *ciénagas*, at least floristically, are far different today than they were in the last century.

There are no pristine riparian wetlands left in southeastern Arizona (Carothers et al. 1974). The major changes have resulted from continuing human impacts—the diversion, damming, and channeling of surface waters and the pumpage of groundwaters so essential to maintaining riparian wetlands. These factors, along with woodcutting, agricultural clearing, construction of transportation corridors, waste disposal, grazing, concentrated human settlement, and a host of other human activities, have made riparian habitats, especially along the primary streams and rivers, the region's most disturbed and degraded habitats.

DESERTSCRUB

Sonoran and Chihuahuan desertscrub dominates the vegetation of the major basin floors and alluvial plains in southeastern Arizona below 4,500 feet in elevation, although Sonoran desertscrub is mostly found on level land below 3,000 feet in elevation. Extensive areas in both vegetation types have been cleared for urban and rural settlement, as well as for irrigated agriculture. Several researchers, such as Hastings and Turner (1965), Martin and Turner (1977), Gehlbach (1981), Goldberg and Turner (1986), and Humphrey (1987), have recorded changes in Sonoran desertscrub in repeat ground photography and permanent plot studies. Except for the extirpation of the native vegetation for settlement, the decline of native grasses, and the invasion of exotics, most of the changes appear to be related to plant life cycles and/or short-term cycles linked to climatic and other environmental fluctuations. For example, Goldberg and Turner (1986) have demonstrated both short- and long-term cyclic changes in the cover, density, and numbers of such native desertscrub dominants as bursage (*Ambrosia deltoidea*), brittlebush (*Encelia farinosa*), white ratany (*Krameria grayi*), *Janusia gracilis*, *Menodora scabra*, staghorn cholla (*Opuntia versicolor*), teddybear cholla (*Opuntia bigelovii*), fairy duster (*Calliandra eriophylla*), *Lycium* spp., and even long-lived species such as ironwood (*Olneya tesota*), foothills paloverde (*Cercidium microphyllum*), creosote bush (*Larrea tridentata*), sahuaro (*Carnegiea gigantea*), and blue paloverde (*Cercidium floridium*).

I have found no evidence that desertscrub communities have invaded extensive areas of former grassland in southeastern Arizona or that the distribution of Chihuahuan and Sonoran desertscrub communities has changed during the historic period. Humphrey (1987) notes that many areas of desertscrub that supported grass or a grass-scrub mixture up to 1893 now support only scrub. He attributes this change to long-exerted grazing pressures and a slight, but consistent, trend toward increased aridity (1987:429).

Although wildfires were never frequent in desertscrub, they were probably more common in the nineteenth century than they are today (Bahre 1985). That wildfires were never very frequent in desertscrub, however, is attested to by the high mortality of succulents at the Granite burn near Florence (McLaughlin and Bowers 1982).

The major directional changes in desertscrub, outside of its extirpation in areas of settlement, seem to be an increase in woody xerophytes such as mesquite, the rapid expansion of exotics, and a general decrease in native grass cover.

GRASSLANDS

Probably the most dramatic change historically in the grasslands, outside of their removal for urban and rural settlement, has been a major decline in grass growth and an increase in woody trees and shrubs, such as mesquite, acacia, one-seed juniper, burroweed, and snakeweed.

I can find no conclusive evidence that the areal extent of the grasslands has changed; that the grasslands have been replaced by Sonoran or Chihuahuan desertscrub; or that mesquite, acacia, or one-seed juniper have extended their ranges in southeastern Arizona since 1870. In every set of matched, wide-angle landscape photographs in which mesquite and acacia now dominate, these same species can usually be identified in the earlier photographs, albeit in much smaller numbers. Malin (1953) and Johnston (1963), both of whom examined the record of mesquite increases throughout the American Southwest, conclude that mesquite has not extended its range but only increased within its range. Brown (1982) and others point out that the semidesert grassland is potentially a perennial grass-shrub–dominated landscape, positioned between desertscrub below and evergreen woodland, chaparral, and plains grassland above, and has always contained these woody "invaders." Nevertheless, the notion persists among some ecologists that mesquite, acacia, one-seed juniper, and other woody xerophytes have recently invaded the grasslands (Parker and Martin 1952; Humphrey 1958, 1987; Buffington and Herbel 1965).

The rapid and extensive modification of the grasslands, not only in southeastern Arizona but all over the American Southwest, followed major Anglo-American settlement and the start of extensive cattle ranching (Thornber 1910; Leopold 1924; Parker and Martin 1952; Humphrey 1958, 1987; Harris 1966; Wright 1980; Archer et al. 1988). Areas of former grassland now dominated by scrubby trees and shrubs in southeastern Arizona include the lower slopes of the Santa Rita Mountains (Humphrey 1958) (see Figure 4.11), the Aravaipa Valley (see Figure 4.8), the area from the foot of the Sierra San Jose to Greenbush Draw on the east side of the San Pedro Valley near the international boundary

(Hastings 1959:65), the Croton Springs region southwest of Willcox (ibid.), the east-facing alluvial fans of the Huachuca and Whetstone mountains (Robert 1869; Rodgers 1965), the plains east and west of the Dragoon and Chiricahua mountains (Robert 1869), and large sections of the San Simon Valley (Hinton 1878) (see Figure 6.3). The mechanisms by which these brush increases have taken place are not well understood, but the increases are generally attributed to a combination of overgrazing and wildfire exclusion (Wright 1980).

Other factors that have resulted in changes in the grasslands are agricultural clearing, wild hay cutting, urban and rural development (especially the expansion of rural subdivisions), range management policies, and the introduction of exotics.

To sum up, the major directional changes in the grasslands have been a dramatic increase in brush and scrubby trees in certain areas, a decline in native grasses, rapid expansion of exotics, and clearing for urban and rural development.

EVERGREEN WOODLANDS

Using twenty-six repeat ground photographs of fifteen different sites in southern Arizona and northern Sonora, Mexico, Hastings and Turner (1965:104) note three general trends in the evergreen woodlands since 1900: (1) oaks are dying faster than they are becoming established, especially at elevations below 4,500 feet; (2) the boundary separating evergreen woodland from grassland is migrating upward throughout southern Arizona; and (3) there has been an invasion of the evergreen woodland by plants "which, if present at all at the time of the early photographs, were less abundant." They attribute these shifts to climatic change or to a trend toward greater aridity.

Although the density of evergreen oaks and understory shrubs has markedly increased within the Coronado National Forest, little or no change has occurred in the areal extent of the evergreen woodlands of southeastern Arizona, except where oaks have been cleared, heavy grazing has occurred, or fuelwood has been cut. In my review of vegetation changes, using the General Land Office surveyors' field notes, oaks are still found in greater or lesser numbers along all of the twenty-three section lines where they were observed by the first surveyors. Of the twenty-two matched transects in evergreen woodlands in my repeat aerial photographs, trees, mostly oaks, had increased along nine transects and decreased along thirteen. Moreover, I can find no changes in the patterning of the evergreen woodlands between the 1935–1937 and

1983–1984 images, except in cleared areas (see Figures 4.13 and 4.14).

This is noteworthy because Hastings and Turner (1965:104) indicate that oak decadence and retreat upslope were just as apparent in their matched photographs of evergreen woodland sites where the first photos were taken, in 1925 and 1931. Finally, my analysis of changes in the evergreen woodlands in repeat ground photography indicates that oaks have increased and decreased in different areas, but their range has not changed. This last finding is supported by Humphrey's analysis of changes in matched photographs of the evergreen woodlands along the U.S.–Mexico boundary (1987:429–430).

Hastings and Turner base most of their evidence for the upward retreat of evergreen woodland in southern Arizona and northern Sonora on the complete disappearance of a few scattered oaks at four different sites at or below 4,200 feet in elevation. They remark, however, that oaks still grow near one site (1965:83) and that one oak still remains at another site (1965:104). Similar changes are not evident in the matched photographs of sites with oaks below 4,200 feet in elevation along the international boundary. Indeed, Humphrey (1987:429–430) specifically points out that the evergreen woodlands in the lower elevation areas "seem in general, to have changed little or not at all." This finding parallels that in California of Griffin (1977:408), who concludes that the boundaries of California's oak woodlands have remained relatively stable in historic times.

Comments by early observers suggest that much of the lower-elevation evergreen woodland in southern Arizona had little woody understory before Anglo settlement (Leopold 1924; Humphrey 1958; Pyne 1982). Fires, set by lightning or by Indians, are usually identified as "disturbances" that kept the woody understory in check and maintained the open nature of the oak woodlands. With Anglo-American settlement, however, fire suppression policies and overgrazing diminished the occurrence of wildfires and apparently allowed brush and scrubby trees to increase (ibid.).

Although different species of oaks have distinct sensitivities to fire (Plumb 1980), nearly all of the oaks, especially the evergreen or live oaks, resprout vigorously from the root crown and below-ground bud zone after a fire (Phillips 1912; Longhurst 1956; Griffin 1980). The impact of fire on oak dynamics in southeastern Arizona, however, remains unpublished. In particular, research on fire susceptibilities of oak and mesquite in the low-elevational evergreen woodlands would shed light on the long-term effects of fire and fire exclusion on the dynamics of these two species.

The recent increase in the density of evergreen oaks and woody shrubs on Forest Service lands is thought to have resulted from both fire suppression and tighter controls over grazing and fuelwood cutting. While, on the one hand, overgrazing has led to increases in brush by lessening the occurrence of wildfire, it may have caused, on the other hand, a decline in oak regeneration. For example, the browsing of young oak seedlings and trees, and heavy damage to acorns, by livestock are often cited as the cause of poor oak regeneration.

In California, deciduous oaks, possibly because they are less browse resistant than evergreen oaks, appear to have reproduced poorly since 1940, especially in heavily grazed areas (White 1966; Griffin 1971, 1976, 1977; Fieblekorn 1972). Evergreen oaks have continued to reproduce adequately. Similar problems with oak regeneration apparently are occurring in Arizona (Hastings and Turner 1965:104–105). In California, Longhurst et al. (1979), who studied the effects of sheep and deer on oak regeneration at the Hopland Station, found that in an exclosure protected from grazing for five years there were 554 seedlings per acre, compared with no seedlings in nearby grazed areas. In other studies, however, exclusion of cattle from plots in oak parkland did not result in increased oak regeneration (Duncan and Clawson 1980).

Historic fuelwood cutting rarely resulted in the killing of oaks, especially live oaks (Phillips 1912; Bahre and Hutchinson 1985), but it probably impacted the structure of many juniper, pinyon, and oak stands in the evergreen woodlands. The ecological impacts of historic fuelwood cutting, however, have been largely ignored. This is surprising, because there is no doubt that (1) wildfires in the evergreen woodlands have diminished since Anglo settlement, (2) livestock have grazed the oak woodlands since the early nineteenth century, and (3) local woodlands supplied immense amounts of cordwood for nineteenth-century mining operations, as well as cooking and heating fuel for most of southern Arizona's homes until the 1940s.

The directional changes in evergreen woodland are continuous clearing of native cover due to expanding settlement, invasion of exotics, and an increase in oak in some "protected areas." Because the evergreen woodlands were already badly disturbed when most of the earliest photographs were taken, it is difficult to identify other directional changes.

PONDEROSA PINE AND MIXED-CONIFER FORESTS

The ponderosa pine and mixed-conifer forests of southeastern Arizona have been under the "protection" of the National Forest system since the Santa Rita, Mt. Graham (Pinaleño), Santa Catalina, and Chiricahua National Forest reserves were first established in 1902 (Toumey 1901; Lauver 1938; Baker et al. 1988). Although most of these forests continue to be logged and grazed, exploitation since the 1930s has not been as ruthless or probably as destructive as it was in the late nineteenth century. Then, the stocking and grazing of cattle, sheep, and goats were largely uncontrolled, and large areas of forest were clear-cut for fuelwood and timber in spite of laws against indiscriminate woodcutting and logging on public domain long before the establishment of the reserves. In 1902, for example, 25,000 cattle and 5,000 sheep were grazed in the Santa Rita Mountains (Range Conditions in Arizona 1900–1909), while Royal Kellogg (1902a:504), who surveyed the pine forests of southeastern Arizona at the turn of the century for the Bureau of Forestry, reported that "all of the good timber [in the Huachuca Mountains] was cut out years ago."

Since the advent of large-scale Anglo-American settlement, wildfire exclusion, logging, grazing, and Forest Service management of these activities, along with fuelwood cutting, mining, and the construction of transportation corridors, have had a major effect on the ecology and age structure of the ponderosa pine and mixed-conifer forests of southeastern Arizona. According to Cooper (1960:161), who studied Arizona's ponderosa pine forests:

> The most important change brought about by the white man has been the virtual exclusion of fire from the forests of the Southwest. Under natural conditions, light surface fires, set by lightning or Indians, burned through all parts of the pine forest at regular intervals of 3 to 10 yrs. These fires acted as natural thinning agents and reduced surplus fuel. The reduction of inflammable grass by grazing animals, and an intensive fire prevention program have largely eliminated fire from the woods. The major cause of the present excess of pine production is exclusion of fire.

Historical and dendrochronological evidence clearly indicates that fire has always been a major ecological force in the ponderosa pine and mixed-conifer forests of southeastern Arizona and that the frequency of fire declined drastically after Anglo-American settlement (Leopold 1924; Weaver 1951a; Cooper 1960; Arno and Sneck 1977; Barrows 1978; Pyne 1982; Bahre 1985; Baisan 1988; Swetnam et al. 1989). According to current thought, wildfires before Anglo settlement, started by lightning or humans (it is impossible to determine the proportion caused by each), thinned stands, eliminated young pines and thickets, and kept the ponderosa pine forests open and parklike with an understory of herbs and shrubs (Weaver 1947, 1951b; Arnold 1950; Cooper 1960; Biswell 1972). In addition, patches of mineral soil laid bare by wildfire provided favorable conditions for the establishment of the shade-intolerant ponderosa pine seedlings (Cooper 1960; Wright and Bailey 1982).

The exclusion of fire in the mountain ranges of the American Southwest after Anglo-American settlement has led to dense thickets, even-aged stands, forest stagnation, and more intense fires than ever before (Weaver 1951a, 1955; Cooper 1960). Furthermore, grazing altered the composition and quantity of ground cover vegetation, and prepared the ground for pine reproduction by removing herbaceous competition and exposing the soil.

Robert Russell (1982), in his study of human impacts since the 1880s on the high mountain meadow vegetation of the Chiricahua Mountains, concludes that (1) severe overgrazing of the meadows, especially at the turn of the century, has led to extirpation of several plant species as well as to alterations in plant composition and the spread of alien plants; (2) fire suppression has led to the invasion of the meadows by ponderosa pine and Douglas fir; and (3) the meadows are drying out, possibly because fire suppression and grazing have allowed the surrounding forests to become denser and thereby change local water regimes.

That logging once led to clear-cutting in the Santa Rita, Pinaleño, Santa Catalina, and Huachuca mountains is evident in the historic record, but its extent and impact are unrecorded. For example, not only was most of the "good timber" in the Huachuca Mountains cut out before the turn of the century, but parts of the Chiricahua Mountains (Barfoot, Pinery Canyon, Rock Canyon, Rucker Canyon, and Morse Canyon) were clear-cut to supply mine timbers and construction materials to Bisbee and Tombstone in the 1880s and 1890s (*Arizona Weekly Star* Nov. 12, 1885; Kellogg 1902b; *Bisbee Review* Aug. 8, 1923). In addition, in 1902 Kellogg (1902a:505) noted that much of the available timber in the Graham (Pinaleño) Mountains had been cut but that more could be reached by road building and that "the forest in the Santa Catalinas is nearly intact, because there are no good roads." Both the Santa Catalina and Pinaleño mountains continue to be harvested, albeit at very low levels.

In conclusion, little or no research has been done on the impact of historic logging, fire suppression, and grazing on the ponderosa pine and mixed-conifer forests of southeastern Arizona. Since the region's settlement, the major directional changes in these forest types appear to be fewer large trees; more pine production and denser thickets; less grass; less frequent though more intense fires, especially crown fires; desiccation and invasion of mountain meadows; the spread of exotic plants; and clearing for transportation rights-of-way, recreational facilities, and mines.

11   Summary and Conclusions

Except for the reduction in numbers or elimination of some wild animals since the 1850s and 1860s, it is not clear that all of the changes in the biological environment of southeastern Arizona have been as dramatic or as extensive as some researchers have implied or that some of the suggested vegetation changes have even occurred. It is certain that most of the directional vegetation changes have happened in different places at different times and that the most dramatic and far-reaching of these were initiated in the late nineteenth century. While short-term cyclic changes in the vegetation, due to climatic fluctuations, succession, and the life cycles of individual genera, seem clear, there is no conclusive evidence that natural environmental fluctuations or reputed climatic shifts or anomalies since the 1870s have caused any major directional changes in the vegetation.

As I have pointed out, the major proponents of the hypothesis that a shift toward greater aridity since 1870 is the major cause of recent regional vegetation change base their conclusion largely on three premises: (1) They reason that wildfire exclusion was probably not a major factor for woody plant increases in the grasslands at the turn of the century because they found no evidence for frequent wildfires in southeastern Arizona during that period. (2) They question whether overgrazing in the nineteenth century could have been the initiating factor for the woody plant increases and stream entrenching because, they say, there were just as many cattle in southeastern Arizona in the 1820s as in the 1890s and these earlier numbers did not cause similar changes. (3) They point out that their repeat photography clearly demonstrates an upward retreat of plant ranges along a xeric-to-mesic gradient.

The evidence examined here, however, contradicts these premises. (1) Wildfires were common throughout southeastern Arizona in the past. (2) There is no solid evidence of the occurrence of large numbers of cattle or extensive overgrazing during Mexican occupation. That cattle numbers during the 1820s were as great as in the 1890s is unlikely because of the essential lack of developed water sources for cattle during the earlier period, the small amount of rangeland within reach of perennial water supplies, and the onset of Apache depredations shortly after the establishment of the large Mexican land grants for cattle ranching. (3) Most important, there is no persuasive evidence in the surveyors' field notes, permanent plot studies, or repeat ground and aerial photography to suggest any changes in the geographic range of the major vegetation types or communities since the 1870s.

Unfortunately, most of the historic ground photographs from which the most accurate information on site-specific changes in the vegetation of southeastern Arizona has been obtained were taken after 1890, long after some of the most striking changes due to Anglo settlement had occurred. For example, only thirteen of the ninety-seven photographic pairs in The Changing Mile (Hastings and Turner 1965) were taken before 1890 (the earliest two in 1883), and these were of townsites or of otherwise heavily disturbed sites. The first photographs of the markers along the Arizona-Sonora boundary were taken during the severe drought of 1891–1893, when thousands of range cattle in southeastern Arizona were dying of thirst and starvation. By 1890 the woodlands of southeastern Arizona had been severely cut over for cordwood, and range cattle were near the highest numbers they would ever be. These two facts, combined with the destruction of much of the riparian wetlands, fire control or at least fire reduction, and the increase of woody plants in certain rangelands, point to an already badly disturbed environment when the earliest landscape photographs were taken. Consequently, in our studies of vegetation change in repeat ground and aerial photography, we are largely assessing short-term fluctuations and directional trends in an already disturbed landscape. Moreover, the surveyors' field notes and the nineteenth-century landscape descriptions are often too vague to evaluate site-specific vegetation changes, let alone regional changes.

The most apparent directional vegetation changes in southeastern Arizona have been the introduction and expansion of exotics; removal of native plant cover for settlement; an overall decline in native grasses; an increase in woody plants, especially in the grasslands and along the lower edge of the evergreen woodlands; an increase in oak, juniper, and ponderosa pine in protected areas; and the general degradation of vegetation cover due to various human activities. All of the changes follow and appear to be due to catastrophic as well as continuing human disturbances.

That the woody plant increase in the rangelands is a result of increasing aridity seems only a remote possibility. Surely it would be difficult to argue that the increase in oaks and ponderosa pine thickets is due to increasing aridity. It is also clearly evident in repeat photography that mesquite and acacia rapidly colonize badly degraded or once-cleared sites throughout southeastern Arizona below 5,000 feet in elevation.

The present historic evidence for anthropogenic vegetation change casts serious doubt on the hypothesis that a shift toward greater aridity is the primary factor for regional vegetation changes since 1870. In particular, two major regional vegetation changes—the upward shift in the range of the major vegetation types and the increase of woody plants in the rangelands—are ascribed to a trend toward increasing aridity. The former change is unproved, and the woody plant increase is most likely the result of grazing and fire exclusion.

Even though few, if any, areas in the United States have escaped the transforming hand of humans, few ecologists have identified and analyzed the anthropogenic factors that have influenced the development of the present vegetation cover, let alone evaluated whether the changes are short-term or directional, or whether they reflect past or continuing disturbances. For the most part, knowledge of land-use dynamics is the weakest link in studies of historic vegetation change not only in the United States but also in the rest of the world.

It is hoped that this book will persuade ecologists to devise methods for undertaking the difficult task of relating historic land-use activities to site-specific vegetation changes, especially in matched aerial and ground photography. Clearly, we must analyze the human land-use activities that have affected the wild landscape historically and ecologically, and we must attempt to understand the mechanisms of vegetation change if we wish to arrive at a better understanding of vegetation dynamics and recovery from human disturbance not only in southeastern Arizona but also in other parts of the world. Any attempt at understanding vegetation dynamics must begin with the assumption that the landscape has been disturbed until it can be demonstrated that the assumption is false. To do otherwise is unwittingly to accept a hypothesis that is usually untrue and can only lead to biased results and a presumed understanding of process that may be wrong.

Appendix A

Selected Section Lines with Oaks Recorded
by the Public Land Survey in Southeastern Arizona

Section-line, Section, Township, and Range	Date of Survey
1. Western, S27, T18S, R16E	1904
2. Western, S34, T22S, R15E	1876
3. Western, S25, T21S, R14E	1887
4. Western, S19, T19S, R16E	1874
5. Western, S13, T20S, R14E	1923
6. Western, S6, T21S, R17E	1912
7. Western, S36, T20S, R16E	1885
8. Western, S26, T20S, R16E	1885
9. Western, S23, T21S, R14E	1887
10. Western, S25, T17S, R23E	1902
11. Western, S15, T22S, R16E	1883
12. Western, S15, T19S, R16E	1874
13. Western, S7, T24S, R17E	1904
14. Western, S20, T23S, R21E	1901
15. Western, S20, T17S, R31E	1884
16. Western, S17, T18S, R24E	1898
17. Western, S23, T18S, R23E	1936
18. Western, S1, T23S, R20E	1901
19. Western, S24, T23S, R20E	1906
20. Western, S14, T23S, R20E	1906
21. Western, S11, T23S, R19E	1927
22. Western, S32, T21S, R16E	1876
23. Northern, S34, T19S, R14E	1923

Appendix B

Tree and Large Woody Shrub Counts Along 111 Matching Transects on SCS (1935–1937) and NHAP (1983–1984) Imagery by Vegetation Type

Vegetation Type	SCS Count	NHAP Count	Absolute Difference	Percent Difference
Grassland	14	10	−4	−29
Grassland	17	8	−9	−53
Grassland	70	67	−3	−4
Grassland	90	93	3	3
Grassland	44	48	4	9
Grassland	4	9	5	125
Grassland	0	0	0	0
Grassland	107	78	−29	−27
Grassland	109	131	22	20
Grassland	60	55	−5	−8
Grassland	37	18	−19	−51
Grassland	79	78	−1	−1
Grassland	35	30	−5	−14
Grassland	34	32	−2	−6
Grassland	50	49	−1	−2
Grassland	26	26	0	0
Grassland	33	32	−1	−3
Grassland	37	34	−3	−8
Grassland	30	38	8	27
Grassland	90	84	−6	−7
Grassland	83	99	16	19
Grassland	122	101	−21	−17
Grassland	86	80	−6	−7
Grassland	116	127	11	9
Grassland	69	77	8	12
Grassland	0	0	0	0
Grassland	2	0	−2	−100

Appendix B (continued)

Vegetation Type	SCS Count	NHAP Count	Absolute Difference	Percent Difference
Grassland	118	128	10	8
Grassland	112	51	−61	−54
Grassland	3	2	−1	−33
Grassland	60	63	3	5
Grassland	60	64	4	7
Grassland	87	63	−24	−28
Grassland	76	74	−2	−3
Grassland	22	48	26	118
Grassland	30	29	−1	−3
Grassland	160	181	21	13
Grassland	106	92	−14	−13
Grassland	84	96	12	14
Grassland	50	78	28	56
Grassland	54	38	−16	−30
Grassland	67	65	−2	−3
Grassland	135	140	5	4
Grassland	101	104	3	3
Grassland	99	60	−39	−39
Grassland	96	71	−25	−26
Grassland	105	149	44	42
Grassland	66	61	−5	−8
Grassland	74	80	6	8
Grassland	130	112	−18	−14
Evergreen Woodland	97	86	−11	−11
Evergreen Woodland	106	92	−14	−13
Evergreen Woodland	188	152	−36	−19
Evergreen Woodland	174	130	−44	−25
Evergreen Woodland	76	61	−15	−20
Evergreen Woodland	63	114	51	81
Evergreen Woodland	90	119	29	32
Evergreen Woodland	83	79	−4	−5
Evergreen Woodland	117	95	−22	−19
Evergreen Woodland	261	156	−105	−40
Evergreen Woodland	280	136	−144	−51
Evergreen Woodland	119	129	10	8
Evergreen Woodland	110	105	−5	−5
Evergreen Woodland	165	169	4	2
Evergreen Woodland	240	246	6	2
Evergreen Woodland	230	230	0	0
Evergreen Woodland	77	81	4	5
Evergreen Woodland	105	73	−32	−30
Evergreen Woodland	78	77	−1	−1
Evergreen Woodland	380	220	−160	−42

Appendix B (continued)

Vegetation Type	SCS Count	NHAP Count	Absolute Difference	Percent Difference
Evergreen Woodland	380	220	−160	−42
Evergreen Woodland	190	224	34	18
Evergreen Woodland	233	274	41	18
Ponderosa Pine and Mixed-conifer Forest	203	210	7	3
Sonoran Desertscrub	241	219	−22	−9
Sonoran Desertscrub	214	193	−21	−10
Sonoran Desertscrub	73	66	−7	−10
Sonoran Desertscrub	55	52	−3	−5
Sonoran Desertscrub	273	271	−2	−1
Sonoran Desertscrub	205	144	−61	−30
Sonoran Desertscrub	150	122	−28	−19
Sonoran Desertscrub	132	72	−60	−45
Sonoran Desertscrub	128	60	−68	−53
Sonoran Desertscrub	200	117	−83	−42
Sonoran Desertscrub	154	57	−97	−63
Sonoran Desertscrub	150	84	−66	−44
Sonoran Desertscrub	136	133	−3	−2
Sonoran Desertscrub	166	90	−76	−46
Sonoran Desertscrub	183	239	56	31
Sonoran Desertscrub	152	167	15	10
Sonoran Desertscrub	127	144	17	13
Sonoran Desertscrub	229	235	6	3
Sonoran Desertscrub	189	296	107	57
Sonoran Desertscrub	156	172	16	10
Sonoran Desertscrub	115	117	2	2
Sonoran Desertscrub	135	130	−5	−4
Sonoran Desertscrub	112	115	3	3
Chihuahuan Desertscrub	142	83	−59	−42
Chihuahuan Desertscrub	148	97	−51	−34
Chihuahuan Desertscrub	125	113	−12	−10
Chihuahuan Desertscrub	92	66	−26	−28
Chihuahuan Desertscrub	57	54	−3	−5
Chihuahuan Desertscrub	46	43	−3	−7
Chihuahuan Desertscrub	8	11	3	38
Chihuahuan Desertscrub	4	5	1	25
Chihuahuan Desertscrub	116	79	−37	−32
Chihuahuan Desertscrub	92	74	−18	−20
Chihuahuan Desertscrub	106	102	−4	−4
Chihuahuan Desertscrub	110	106	−4	−4
Chihuahuan Desertscrub	31	26	−5	−16
Chihuahuan Desertscrub	48	37	−11	−23
Chihuahuan Desertscrub	93	80	−13	−14

Appendix C

Z-Scores for the Total Sample

Vegetation Type	Decreasing Ranks		Increasing Ranks	
	No.	Mean	No.	Mean
Total Sample	37	47.905	69	56.5
Grassland	19	23.816	28	24.125
Sonoran Desertscrub	8	9.25	15	13.467
Evergreen Woodland	8	9.25	13	12.077
Chihuahuan Desertscrub	2	10	13	7.692
Total Sample Less Chihuahuan Desertscrub	35	40.957	56	49.152
Total Sample Less Chihuahuan and Sonoran Desertscrub	27	32.13	41	36.061
Chihuahuan and Sonoran Desertscrub	10	16	28	20.75

Cases with No Difference	Z-Score	Z-Score Corrected for Ties	No. of Tied Groups
4	−3.35	−3.35	2
3	−1.18	−1.18	1
0	−1.947	0	0
1	−1.442	0	0
0	−2.272	0	0
4	−2.61	−2.61	2
4	−1.867	−1.867	2
0	−3.053	0	0

Notes

CHAPTER 1. INTRODUCTION

1. Studies of human impacts on the environment date from at least the eighteenth-century works of Georges-Louis Leclerc, Comte de Buffon, in his classic *Histoire naturelle*, published in forty-four volumes between 1749 and 1804. For a review of this theme in geography, see Glacken (1960, 1967), Dickinson (1969), Stott (1984), and Kates (1987).

2. For studies by Berkeley-trained geographers dealing with human impacts on vegetation and vegetation change, see Clark (1949), Parsons (1955, 1981), Aschmann (1956), Burcham (1957), Denevan (1961), Johannessen (1963), Harris (1965, 1966), Smith (1981), Vale (1982), and Veblen and Steward (1982).

CHAPTER 2. SETTING

1. For detailed descriptions of the climate of southeastern Arizona, see Sellers (1960), Hastings and Turner (1965), Sellers and Hill (1974), Cooke and Reeves (1976), and Balling (1988, 1989).

2. The popular biotic classifications in parentheses are used by Brown and Lowe (1980) and Brown (1982) to describe the vegetation cover of southeastern Arizona.

3. The percentages of cover for each major vegetation type were estimated from Brown and Lowe (1980).

4. Plant nomenclature follows that of Lehr in *A Catalogue of the Flora of Arizona* (1978).

5. Early descriptions of the landscape of southeastern Arizona are reviewed in Hastings (1959), Hastings and Turner (1965), Dobyns (1981), and Hendrickson and Minckley (1984).

CHAPTER 5. LIVESTOCK GRAZING

1. See also Humphrey (1987), who has reproduced all of the original photographs of the boundary markers between the Rio Grande and the Colorado River.

2. For information on cattle impacts on riparian habitats in southeastern Arizona, see Johnson and Jones (1977) and Johnson et al. (1985).

CHAPTER 7. FUELWOOD CUTTING

1. On these uses of wood and the types of wood preferred for fuel, see *Tombstone Epitaph* (Aug. 27, 1880), *Arizona Daily Star* (Oct. 24, 1880), Austin

(1883), and Peters (1911:385, 404). Like their nineteenth-century counterparts, modern Mexican and Tohono O'odham Indian woodcutters claim that trees smaller than six inches in diameter are not worth the time to cut down and clean if there are larger trees in the stand. Depending on the size and spacing of the trees and the nature of the terrain, a good woodcutter can usually cut a cord of wood per day in the mesquite and evergreen woodlands.

2. For descriptions of woodcutting and woodcutters, see Miller (n.d.:2, 9), *Arizona Daily Star* (Dec. 14, 1880; Mar. 24, 1882; May 13, 1882; Mar. 28, 1883), *Tombstone Daily Nugget* (Feb. 5, 1882), *Daily Tombstone* (June 2, 1886), Rockfellow (1955:108), Dobyns (1981), *Arizona Silver Belt* (Aug. 23, 1890).

3. Most of this section on Tombstone is from Bahre and Hutchinson (1985).

4. Cordage figures for these mountain ranges were obtained from several sources. I consulted the following typed reports in the Douglas, Arizona, office of the Coronado National Forest: Houck and Ambrose (1978), Weeden (1980, 1981). In the Sierra Vista, Arizona, office of the Coronado National Forest, I read Van Sickle and Newman (1981). The figures also reflect an interview with Lee Poague, Coronado National Forest, Tucson, Arizona, in June 1983.

5. The amounts of fuelwood required to mill each ton of ore for all of the stamp mills in the district ranged from 0.11 cord to 0.16 cord. See Staunton (n.d.), Austin (1883:93), and Church (1887).

6. See the *Arizona Daily Star* (Mar. 1, 1882; Apr. 23, 1884) and the U.S. General Land Office surveyors' field notes (1885–1916) for the Tombstone District at the Bureau of Land Management in Phoenix, Arizona.

CHAPTER 8. EXOTIC PLANT INTRODUCTIONS

1. For instance, my rough count of exotic plants in Hendrickson and Minckley's flora of *ciénaga* habitats in southeastern Arizona (1984) indicates that approximately 23 percent of the modern *ciénaga* flora is made up of exotics.

2. See Burgess et al. (in press) and Schmutz et al. (1968) for brief histories of several exotics in southern Arizona.

CHAPTER 9. AGRICULTURE, LOGGING, AND HAYING

1. For more information on major floods in southeastern Arizona and the role of agricultural disturbance in initiating stream entrenching, see Greeley and Glassford (1891), Minckley and Clark (1984:28), Smith and Hackler (1955), Cooke and Reeves (1976), Dobyns (1981), Waters (1988), and Betancourt (1990).

2. The formula for determining the board foot content of sawed lumber is

$$\text{bd. ft.} = \frac{\text{thickness (in.)} \times \text{width (in.)} \times \text{length (ft.)}}{12}.$$

3. Maps showing logged areas in the Pinaleños between 1903 and 1925 are available in the library of the Coronado National Forest regional office in Tucson.

Bibliography

Acocks, J. P. H. 1953. Veld Types of South Africa. *Memoirs of the Botanical Survey of South Africa* 28:1–192.
Agee, J. K., and P. Dunwiddie. 1984. Recent Forest Development on Yellow Island, San Juan County, WA. *Canadian Journal of Forestry* 62:2074–2080.
Ahlcrona, E. 1988. *The Impact of Climate and Man on Land Transformation in Central Sudan, Applications of Remote Sensing*. Meddelanden från Lunds Universitets Geografiska Institutioner, avhandlingar 103. Lund, Sweden: Royal University of Lund.
Ahlstrand, G. M. 1980. Fire History of a Mixed Conifer Forest in the Guadalupe Mountains National Park. In *Proceedings of the Fire History Workshop, October 20–24, 1980. Tucson, Arizona*, 4–7. USDA Forest Service General Technical Report RM-81. Fort Collins, Colo.: Rocky Mountain Forest and Range Experiment Station.
Alexander, R. R. 1974. *Silviculture of Central and Southern Rocky Mountain Forests: A Summary of the Status of Our Knowledge by Timber Types*. USDA Forest Service Research Paper RM-120. Fort Collins, Colo.: Rocky Mountain Forest and Range Experiment Station.
Allen, L. S. 1989. Livestock and the Coronado National Forest. *Rangelands* 11:14–20.
Ames, C. R. 1977a. Along the Mexican Border—Then and Now. *Journal of Arizona History* 18:431–446.
———. 1977b. Wildlife Conflicts in Riparian Management: Grazing. In *Importance, Preservation, and Management of Riparian Habitats*, edited by R. R. Johnson and D. R. Jones, 49–51. USDA Forest Service General Technical Report RM-43. Fort Collins, Colo.: Rocky Mountain Forest and Range Experiment Station.
Antevs, E. 1952. Arroyo-Cutting and Filling. *Journal of Geology* 60:375–385.
———. 1962. Late Quaternary Climates in Arizona. *American Antiquity* 28:193–198.
Applegate, L. H. 1981. Hydraulic Effects of Vegetation Changes Along the Santa Cruz River near Tumacacori, Arizona. Master's thesis, University of Arizona.
Archer, C., C. Scifres, and C. R. Bassham. 1988. Autogenic Succession in a Subtropical Savanna: Conversion of Grassland to Thorn Woodland. *Ecological Monographs* 58:111–127.

198   Bibliography

Arizona Agriculture: Now and a Vision of the Future. 1986. College of Agriculture, University of Arizona.

Arizona Crop and Livestock Reporting Service. 1966. *Arizona Agricultural Statistics, 1867 to 1965*. Phoenix.

———. 1985. *1984 Arizona Agricultural Statistics*. Bulletin S-19. Phoenix: USDA and University of Arizona.

Arizona Daily Star (Tucson), July 30, 1879; Aug. 31, 1879; Feb. 10, 1880; June 2, 1880; June 3, 1880; Sept. 2, 1880; Oct. 24, 1880; Dec. 14, 1880; Feb. 23, 1882; Feb. 24, 1882; Mar. 1, 1882; Mar. 17, 1882; Mar. 24, 1882; Apr. 16, 1882; May 13, 1882; May 16, 1882; May 21, 1882; June 27, 1882; July 19, 1882; Sept. 27, 1882; Nov. 12, 1882; Nov. 29, 1882; Mar. 28, 1883; June 9, 1883; June 23, 1883; Mar. 7, 1884; Apr. 23, 1884; Oct. 22, 1884; Feb. 17, 1885; June 13, 1886; July 7, 1886; May 17, 1887; June 9, 1887; June 14, 1887; June 15, 1887; Dec. 14, 1887; June 19, 1888; June 22, 1889; Oct. 12, 1889; Oct. 24, 1889; Jan. 5, 1890; Feb. 12, 1890; Mar. 20, 1892; Mar. 29, 1892; May 28, 1892; Dec. 20, 1892; Oct. 14, 1902; Aug. 18, 1904; Sept. 8, 1905; Nov. 11, 1987.

Arizona Silver Belt (Globe), May 30, 1879; May 1, 1880; June 12, 1880; Aug. 23, 1890.

Arizona Star (Tucson), June 21, 1877; June 23, 1877. (In 1879 the name was changed to *Arizona Daily Star*.)

Arizona Weekly Citizen (Tucson), July 5, 1873; July 21, 1873; Sept. 20, 1873; Nov. 8, 1873; Dec. 6, 1873; May 2, 1874; June 6, 1874; June 13, 1874; June 20, 1874; Dec. 5, 1874; Mar. 13, 1875; July 1, 1876; June 30, 1877; Aug. 4, 1877; Oct. 26, 1878; Aug. 1, 1879; Oct. 25, 1879; May 8, 1880; June 2, 1883; June 30, 1883.

Arizona Weekly Enterprise (Florence), Feb. 18, 1882; Aug. 7, 1886; Mar. 17, 1888; Aug. 10, 1889; Apr. 5, 1890.

Arizona Weekly Star (Tucson), May 23, 1878; Jan. 1, 1879; May 22, 1879; June 5, 1879; June 19, 1879; June 26, 1879; May 27, 1880; Nov. 11, 1883; Nov. 12, 1885.

Arizona: The New State Magazine (Phoenix). 1912. 3:5.

Arno, S. F., and K. M. Sneck. 1977. *A Method for Determining Fire History in Coniferous Forests in the Mountain West*. USDA Forest Service General Technical Report INF-42. Ogden, Utah: Intermountain Forest and Range Experiment Station.

Arnold, J. F. 1950. Changes in Ponderosa Pine Bunch-grass Ranges in Northern Arizona Resulting from Pine Regeneration and Grazing. *Journal of Forestry* 48:118–126.

Arnold, J. F., D. A. Jameson, and E. H. Reid. 1964. *The Pinyon-Juniper Type of Arizona: Effects of Grazing, Fire, and Tree Control*. USDA Forest Service Production Research Report 84. Washington, D.C.: GPO.

Artesian Belt—San Simon, Arizona—Homeseekers' Edition. 1914. 3(47).

Aschmann, H. H. 1956. The Evolution of a Wild Landscape and Its Persistence in Southern California. *Annals of the Association of American Geographers* 49:34–56.

———. 1970. Athapaskan Expansion in the Southwest. *Yearbook, Association of Pacific Coast Geographers* 32:79–97.

Austin, L. W. 1883. Silver Mining in Arizona. *Transactions of the American Institute of Mining Engineers* 11:91–106.

Bahre, C. J. 1977. Land-Use History of the Research Ranch, Elgin, Arizona. *Journal of the Arizona Academy of Science* 12 (supp. 2): 1–32.

———. 1979. *Destruction of the Natural Vegetation of North-Central Chile.* University of California Publications in Geography, vol. 23. Berkeley: University of California Press.

———. 1984. Effects of Historic Fuelwood Cutting on the Semidesert Woodlands of the Arizona-Sonora Borderlands. In *History of Sustained-Yield Forestry: A Symposium*, edited by H. K. Steen, 101–110. Portland, Ore.: Forest History Society of America.

———. 1985. Wildfire in Southeastern Arizona Between 1859 and 1890. *Desert Plants* 7:190–194.

———. 1987. Wild Hay Harvesting in Southern Arizona: A Casualty of the March of Progress. *Journal of Arizona History* 28:69–78.

Bahre, C. J., and D. E. Bradbury. 1978. Vegetation Change Along the Arizona-Sonora Boundary. *Annals of the Association of American Geographers* 68:145–165.

Bahre, C. J., and C. F. Hutchinson. 1985. The Impact of Historic Fuelwood Cutting on the Semidesert Woodlands of Southeastern Arizona. *Journal of Forest History* 29:175–186.

Bailey, R. W. 1935. Epicycles of Erosion in the Valleys of the Colorado Plateau Province. *Journal of Geology* 63:337–355.

Baisan, C. H. 1988. Fire History of the Rincon Mountain Wilderness, Saguaro National Monument. Unpublished manuscript prepared for Saguaro National Monument, Tucson.

Baker, R. D., R. S. Maxwell, V. H. Treat, and H. C. Dethloff. 1988. *Timeless Heritage: A History of the Forest Service in the Southwest.* USDA Forest Service FS-409. Washington, D.C.: GPO.

Balling, R. C. 1988. The Climatic Impact of a Sonoran Vegetation Discontinuity. *Climatic Change* 13:99–109.

———. 1989. The Impact of Summer Rainfall on the Temperature Gradient Along the United States-Mexico Border. *Journal of Applied Meteorology* 28:304–308.

Bandelier, A. F. 1890. Hemenway Southwestern Archaeological Expedition: Contributions to the History of the Southwestern Portion of the United States. *Papers of the Archeological Institute of America.* American Series, vol. 5. Cambridge, Mass.: John Wilson and Son.

Barnes, W. C. 1935. *Arizona Place Names.* Tucson: University of Arizona Press.

———. 1936. Herds in San Simon Valley. *American Forests* 42:456–457, 481.

Barr, F. R. 1940. The Detroit Copper Mining Company of Arizona, 1875–1919. Unpublished manuscript, Arizona Historical Society, Tucson.

Barrows, J. S. 1978. *Lightning Fires in Southwestern Forests.* Final Report (16-568-CA) prepared by Colorado State University. Fort Collins, Colo.: Intermountain Forest and Range Experiment Station.

Bartlett, J. R. 1854. *Personal Narrative of Explorations and Incidents in Texas, New Mexico, California, Sonora, and Chihuahua Connected with the*

United States and Mexico Boundary Commission During the Years 1850, '51, and '53. 2 vols. New York: D. Appleton & Co.

Beatley, J. C. 1966. Ecological Status of Introduced Brome Grasses (*Bromus* spp.) in Desert Vegetation of Southern Nevada. *Ecology* 47:548–554.

———. 1967. Survival of Winter Annuals in the Northern Mojave Desert. *Ecology* 48:745–750.

Betancourt, J. L. 1986. Paleoecology of Pinyon-Juniper Woodlands: Summary. In *Proceedings—Pinyon Juniper Conference Reno, Nev., June 13–16, 1986,* compiled by R. L. Everett, 129–139. Ogden, Utah: USDA Intermountain Forest and Range Experiment Station.

———. 1990. Tucson's Santa Cruz River and the Arroyo Legacy. Ph.D. diss., University of Arizona.

Bisbee Review, Aug. 8, 1923.

Biswell, H. H. 1956. Ecology of California Grasslands. *Journal of Range Management* 9:19–24.

———. 1972. Fire Ecology in Ponderosa Pine-Grassland. *Proceedings—Tall Timbers Fire Ecology Conference,* no. 12, 69–96. Tallahassee, Fla.: Tall Timbers Research Station.

Blackburn, W. H., and P. T. Tueller. 1970. Pinyon and Juniper Invasion in Black Sagebrush Communities in East-Central Nevada. *Ecology* 51:841–848.

Blake, W. P. 1902. *Tombstone and Its Mines.* New York: Cheltenham Press.

Bock, C. E., and J. H. Bock. 1978. Response of Birds, Small Mammals, and Vegetation to Burning Sacaton Grasslands in Southeastern Arizona. *Journal of Range Management* 31:296–300.

Bock, C. E., J. H. Bock, K. L. Jepson, and J. C. Ortega. 1986. Ecological Effects of Planting African Lovegrasses in Arizona. *Research* 2:456–463.

Bock, J. H., C. E. Bock, and J. R. McKnight. 1976. A Study of the Effects of Grassland Fires at the Research Ranch in Southeastern Arizona. *Journal of the Arizona Academy of Science* 11:49–57.

Bogusch, E. R. 1952. Brush Invasion on the Rio Grande Plains of Texas. *Texas Journal of Science* 4:85–90.

Bolton, H. E., trans. and ed. 1919. *Kino's Historical Memoir of Pimería Alta.* 2 vols. Cleveland: A. H. Clark.

———. 1936. *Rim of Christendom.* New York: Macmillan.

———. 1949. *Coronado, Knight of Pueblos and Plains.* New York: Whittlesey House.

———, ed. 1952. *Spanish Exploration in the Southwest, 1542–1706.* New York: Barnes and Noble.

Bourdo, E. A. 1956. A Review of the General Land Office Survey and of Its Use in Quantitative Studies of Former Forests. *Ecology* 37:754–768.

Bowden, C. 1977. *Killing the Hidden Waters.* Austin: University of Texas Press.

Bowers, J. E. 1981. Catastrophic Freezes in the Sonoran Desert. *Desert Plants* 2:232–236.

Bowers, J. E., and R. M. Turner. 1985. A Revised Vascular Flora of Tumamoc Hill, Tucson, Arizona. *Madroño* 32:225–252.

Brady, P. R. n.d. Ms.89, Brady Collection. Arizona Historical Society, Tucson.

Brand, D. D. 1933. The Historical Geography of Northwestern Chihuahua. Ph.D. diss., University of California, Berkeley.

Brandes, R. 1960. *Frontier Military Posts of Arizona.* Globe, Ariz.: Dale Stuart King.

Brandt, H. 1951. *Arizona and Its Bird Life.* Cleveland: The Bird Research Foundation.

Branscomb, B. L. 1956. Shrub Invasion of a New Mexico Desert Grassland. Master's thesis, University of Arizona.

———. 1958. Shrub Invasion of a Southern New Mexico Desert Grassland Range. *Journal of Range Management* 1:129–132.

Bray, J. R. 1971. Vegetational Distribution, Tree Growth and Crop Success in Relation to Recent Climatic Change. *Advances in Ecological Research* 7:177–233.

Brinckerhoff, S. B. 1967. The Last Years of Spanish Arizona, 1786–1821. *Arizona and the West* 9:5–20.

Brown, A. A., and K. P. Davis. 1973. *Forest Fire: Control and Use.* 2nd ed. New York: McGraw-Hill.

Brown, A. L. 1950. Shrub Invasion of Southern Arizona Desert Grassland. *Journal of Range Management* 3:172–177.

Brown, C. S. 1881. An Arizona Mining District. *The Californian* 5:49–57.

———. 1893. An Adventure in the Huachucas. *Overland Monthly* 21:524–528.

Brown, D. E., ed. 1982. Biotic Communities of the American Southwest: United States and Mexico. *Desert Plants* 4:1–341.

Brown, D. E., N. B. Carmony, and R. M. Turner. 1981. Drainage Map of Arizona Showing Perennial Streams and Some Important Wetlands. Scale 1:1,000,000. 2d ed. Phoenix: Arizona Game and Fish Department.

Brown, D. E., and C. H. Lowe. 1980. *Biotic Communities of the Southwest.* USDA Forest Service General Technical Report RM 78. Map. Fort Collins, Colo.: Rocky Mountain Forest and Range Experiment Station.

Brown, D. E., and R. A. Minnich. 1986. Fire and Changes in Creosote Bush Scrub of the Western Sonoran Desert, California. *American Midland Naturalist* 116:411–422.

Bryan, K. 1922. Erosion and Sedimentation in the Papago Country, Arizona. *U.S. Geological Survey Bulletin* 730-B:1–90.

———. 1925. Date of Channel Trenching (Arroyo Cutting) in the Arid Southwest. *Science* 62:338–344.

———. 1928. Change in Plant Associations by Change in Groundwater Level. *Ecology* 9:474–478.

———. 1940. Erosion in the Valleys of the Southwest. *New Mexico Quarterly* 10:227–232.

Buffington, L. C., and C. H. Herbel. 1965. Vegetational Changes on a Semidesert Grassland Range from 1858 to 1963. *Ecological Monographs* 35:139–164.

Buffon, G. L. L., Comte de. 1749–1804. *Histoire naturelle, générale et particulière.* 44 vols. Paris: L'Imprimerie.

Burcham, L. T. 1957. *California Rangeland: An Historic-Ecological Study of the Range Resources of California.* Sacramento: California Division of Forestry.

Burgess, T. L., R. M. Turner, and J. E. Bowers. (In press). Alien Plants on Tumamoc Hill, Tucson, Arizona, and Their Implications for Long-term Vegetation Change.

Burkham, D. E. 1970. Precipitation, Streamflow, and Major Floods at Selected Sites in the Gila River Drainage Basin Above Coolidge Dam, Arizona. *U.S. Geological Survey Professional Paper* 655-B:1–33.

———. 1972. Channel Changes of the Gila River in Safford Valley, Arizona, 1846–1970. *U.S. Geological Survey Professional Paper* 655-G: 1–24.

———. 1976. Hydraulic Effects of Changes in Bottom-land Vegetation on Three Major Floods, Gila River in Southeastern Arizona. *U.S. Geological Survey Professional Paper* 655-J:1–14.

Burkhardt, J. W., and E. W. Tisdale. 1976. Causes of Juniper Invasion in Southwestern Idaho. *Ecology* 57:472–484.

Cable, D. R. 1961. Small Velvet Mesquite Seedlings Survive Burning. *Journal of Range Management* 14:160–161.

———. 1965. Damage to Mesquite, Lehmann Lovegrass and Black Grama by a Hot June Fire. *Journal of Range Management* 18:326–329.

———. 1967. Fire Effects on Semidesert Grasses and Shrubs. *Journal of Range Management* 20:170–176.

———. 1969. Competition in the Semidesert Grass-Shrub Type as Influenced by Root Systems, Growth Habits, and Soil Moisture Extraction. *Ecology* 50:27–38.

———. 1971. Lehmann Lovegrass in the Santa Rita Experimental Range, 1937–1968. *Journal of Range Management* 24:17–21.

———. 1972. Fire Effects in Southwestern Semidesert Grass-Shrub Communities. *Proceedings—Tall Timbers Fire Ecology Conference*, no. 12, 109–127. Tallahassee, Fla.: Tall Timbers Research Station.

Cable, D. R., and S. C. Martin. 1973. Invasion of Semidesert Grassland by Velvet Mesquite and Associated Vegetation Changes. *Journal of the Arizona Academy of Science* 8:127–134.

Cameron, C. 1896. Report of Colin Cameron, Esq. In U.S. Congress, House, *Annual Report of the Governor of Arizona, 1896.* H. ex. doc. 5, 54th Cong., 2nd sess. Serial 3490, 222–231.

Campbell, G. J., and W. Green. 1968. Perpetual Succession of Stream-Channel Vegetation in a Semiarid Region. *Journal of the Arizona Academy of Science* 5:86–98.

Carmichael, R. S., O. D. Knipe, C. P. Pase, and W. W. Brady. 1978. *Arizona Chaparral: Plant Associations and Ecology.* USDA Forest Service Research Paper RM-202. Fort Collins, Colo.: Rocky Mountain Forest and Range Experiment Station.

Carothers, S. W., R. R. Johnson, and S. W. Aitchison. 1974. Population Structure and Social Organization in Southwestern Riparian Birds. *American Zoologist* 14:97–108.

Carter, M. G. 1964. Effects of Drought on Mesquite. *Journal of Range Management* 17:275–276.

Castetter, E. F. 1956. The Vegetation of New Mexico. *New Mexico Quarterly* 26:257–288.

Cave, G. H. 1982. Ecological Effects of Fire in the Upper Sonoran Desert. Master's thesis, Arizona State University.

Cave, G. H., and D. T. Patten. 1984. Short-term Vegetation Responses to Fire in the Upper Sonoran Desert. *Journal of Range Management* 37:491–495.

Chalmers, M. H. n.d. Reminiscences. Unpublished manuscript, Arizona Historical Society, Tucson.
Christiansen, L. D. 1988. The Extinction of Wild Cattle in Southern Arizona. *Journal of Arizona History* 29:89–100.
Church, J. A. 1887. Concentration and Smelting at Tombstone, Arizona. *Transactions of the American Institute of Mining Engineers* 15:601–613.
Clark, A. H. 1949. *The Invasion of New Zealand by People, Plants and Animals: The South Island.* New Brunswick, N.J.: Rutgers University Press.
Colquhoun, J. 1924. *History of the Clifton-Morenci Mining District.* London: John Murray.
Cook, S. F. 1949. *Soil Erosion and Population in Central Mexico.* Ibero-Americana, vol. 34. Berkeley: University of California Press.
Cooke, P. St. G. 1938. Cooke's Journal of the March of the Mormon Battalion, 1846–1847. In *Exploring Southwest Trails, 1846–1854*, edited by R. P. Bieber and A. B. Bender, 63–240. Southwest Historical Series, vol. 7. Glendale, Calif.: Arthur H. Clark Co.
Cooke, R. U., and R. W. Reeves. 1976. *Arroyos and Environmental Change in the American South-West.* Oxford: Clarendon Press.
Cooper, C. F. 1960. Changes in Vegetation, Structure, and Growth of Southwestern Pine Forests Since White Settlement. *Ecological Monographs* 30:129–164.
Cox, J. R. 1984a. Temperature, Timing of Precipitation and Soil Texture Effects on Germination, Emergence and Seedling Survival of South African Lovegrasses. *Journal of South Africa Botany* 50:159–170.
———. 1984b. Shoot Production and Biomass Transfer of Big Sacaton (*Sporobolus wrightii*). *Journal of Range Management* 37:377–380.
Cox, J. R., R. D. Madrigal, and G. W. Frasier. 1987. Survival of Perennial Grass Transplants in the Sonoran Desert of the Southwestern U.S.A. *Arid Soil Research and Rehabilitation* 1:77–87.
Cox, J. R., and H. L. Morton. 1986. Big Sacaton (*Sporobolus wrightii*) Riparian Grassland Management: Annual Winter Burning, Annual Winter Mowing, and Spring-Summer Grazing. *Applied Agricultural Research* 1:105–111.
Cox, J. R., H. L. Morton, T. N. Johnsen, Jr., G. L. Jordan, S. C. Martin, and L. C. Fierro. 1984. Vegetation Restoration in the Chihuahuan and Sonoran Deserts of North America. *Rangelands* 6:112–116.
Cox, J. R., and G. G. Ruyle. 1986. Influence of Climatic and Edaphic Factors on the Distribution of *Eragrostis lehmanniana* Nees in Arizona, USA. *Journal of the Grassland Society of South Africa* 3:25–29.
Crider, F. J. 1945. *Three Introduced Lovegrasses for Soil Conservation.* USDA Circular 730. Washington, D.C.: GPO.
Culler, R. C., R. L. Hanson, R. M. Myrick, R. M. Turner, and F. P. Kipple. 1982. Evapotranspiration Before and After Clearing Phreatophytes, Gila River Floodplain, Graham County, Arizona. *U.S. Geological Survey Professional Paper* 655-P:1–67.
Daily Tombstone, Apr. 7, 1886; May 12, 1886; June 2, 1886; Nov. 22, 1886.
Darrow, R. A. 1944. Arizona Range Resources and Their Utilization. I. Cochise County. *University of Arizona Agricultural Experiment Station Technical Bulletin* 103:311–366.

Davis, G. P. 1982. *Man and Wildlife in Arizona: The American Exploration Period 1824–1865*. Phoenix: Arizona Game and Fish Department.

Davis, O. K., and R. M. Turner. 1986. Palynological Evidence for the Historic Expansion of Juniper and Desert Shrubs in Arizona, U.S.A. *Review of Palaeobotany and Palynology* 49:177–193.

Denevan, W. M. 1961. *The Upland Pine Forests of Nicaragua: A Study in Cultural Plant Geography*. University of California Publications in Geography, vol. 12. Berkeley: University of California Press.

———. 1967. Livestock Numbers in Nineteenth-Century New Mexico, and the Problem of Gullying in the Southwest. *Annals of the Association of American Geographers* 57:691–703.

Dickinson, R. E. 1969. *The Makers of Modern Geography*. New York: Frederick A. Praeger.

Dieterich, J. H. 1983. Fire History of Southwestern Mixed Conifer: A Case Study. *Forest Ecology and Management* 6:13–31.

DiPeso, C. C. 1951. *The Babocomari Village Site on the Babocomari River, Southeastern Arizona*. Dragoon, Ariz.: Amerind Foundation.

———. 1953. *The Sobaipuri Indians of the Upper San Pedro River Valley, Southeastern Arizona*. Dragoon, Ariz.: Amerind Foundation.

———. 1956. *The Upper Pima of San Cayetano del Tumacácori*. Dragoon, Ariz.: Amerind Foundation.

Dobyns, H. F. 1963. Indian Extinction in the Middle Santa Cruz River Valley, Arizona. *New Mexico Historical Review* 38:163–181.

———. 1978. Who Killed the Gila? *Journal of Arizona History* 19:17–30.

———. 1981. *From Fire to Flood: Historic Human Destruction of Sonoran Desert Riverine Oases*. Ballena Press Anthropology Papers, no. 20. Socorro, N. M.: Ballena Press.

Dodge, M. 1972. Forest Fuel Accumulation—A Growing Problem. *Science* 177:139–142.

Duncan, D. A., and W. J. Clawson. 1980. Livestock Utilization of California's Oak Woodlands. In *Proceedings of the Symposium on the Ecology, Management, and Utilization of California Oaks (June 26–28, 1979, in Claremont, Calif.)*, coordinated by T. R. Plumb, 306–313. USDA Forest Service General Technical Report PSW-44. Berkeley, Calif.: Pacific Southwest Forest and Range Experiment Station.

Emory, W. H. 1857. *Report on the United States and Mexican Boundary Survey*. 2 vols. S. ex. doc. 108, 34th Cong., 1st sess. Washington, D.C.: A. O. P. Nicholson.

Euler, R. C., G. J. Gumerman, T. N. V. Karlstrom, J. S. Dean, and R. H. Hevly. 1979. The Colorado Plateau: Cultural Dynamics and Paleoenvironment. *Science* 205:1089–1101.

Everitt, B. L. 1968. Use of the Cottonwood in an Investigation of the Recent History of a Floodplain. *American Journal of Science* 266:417–439.

Ffolliott, P. F., W. O. Rasmussen, T. K. Warfield, and D. S. Borland. 1979. *Supply, Demand, and Economics of Fuelwood Markets in Selected Population Centers of Arizona*. Arizona Land Marks, vol. 9. Flagstaff: Arizona State Land Department.

Fieblekorn, C. 1972. Interim Report of Oak Regeneration Study. Unpublished report on file, Natural Resources Conservation Office, Hunter Liggett Military Reservation, Jolon, Calif.

Forbes, R. H. 1902. *The River-Irrigating Waters of Arizona—Their Character and Effects.* University of Arizona Agricultural Experiment Station Bulletin 44.

———. 1911. *Irrigation and Agricultural Practice in Arizona.* University of Arizona Agricultural Experiment Station Bulletin 63.

Fraizer, C. n.d. Reminiscences. Tape in Museum of Pimería Alta, Nogales, Ariz.

Frasier, G. W., D. A. Woolhiser, and J. R. Cox, Jr. 1984. Emergence and Seedling Survival of Two Warm-Season Grasses as Influenced by the Timing of Precipitation: A Greenhouse Study. *Journal of Range Management* 37:7–11.

Frederick, K. D. 1982. *Water for Western Agriculture.* Washington, D.C.: Resources for the Future.

Freeman, D. 1979. Lehmann Lovegrass. *Rangelands* 1:162–163.

Frenkel, R. E. 1970. *Ruderal Vegetation Along Some California Roadsides.* University of California Publications in Geography, vol. 20. Berkeley: University of California Press.

Gardner, J. L. 1951. Vegetation of the Creosote Bush Area of the Rio Grande Valley in New Mexico. *Ecological Monographs* 21:379–403.

Gehlbach, F. R. 1981. *Mountain Islands and Desert Seas.* College Station: Texas A&M University Press.

Gerald, R. E. 1968. *Spanish Presidios of the Late Eighteenth Century in Northern New Spain.* Museum of New Mexico Research Records, no. 7. Albuquerque: Museum of New Mexico Press.

Gird, R. 1907. True Story of the Discovery of Tombstone. *Out West* 27:39–50.

Glacken, C. J. 1960. Count Buffon on Cultural Changes of the Physical Environment. *Annals of the Association of American Geographers* 50:1–21.

———. 1967. *Traces on the Rhodian Shore.* Berkeley: University of California Press.

Glendening, G. E. 1952. Some Quantitative Data on the Increase of Mesquite and Cactus on a Desert Grassland Range in Southern Arizona. *Ecology* 33:319–328.

Glendening, G. E., and D. E. Brown. 1982. Mesquite (*Prosopis juliflora*) Response to Severe Freezing in Southeastern Arizona. *Journal of the Arizona-Nevada Academy of Science* 17:15–18.

Glendening, G. E., and H. A. Paulsen, Jr. 1955. *Reproduction and Establishment of Velvet Mesquite as Related to Invasion of Semidesert Grasslands.* USDA Forest Service Technical Bulletin 1127. Washington, D.C.: GPO.

Glinski, R. L. 1977. Regeneration and Distribution of Sycamore and Cottonwood Trees Along Sonoita Creek, Santa Cruz County, Arizona. In *Importance, Preservation and Management of Riparian Habitat: A Symposium,* coordinated by R. R. Johnson and D. A. Jones, 116–123. USDA Forest Service General Technical Report RM-43. Fort Collins, Colo.: Rocky Mountain Forest and Range Experiment Station.

Goetzmann, W. H. 1959. *Army Exploration in the American West, 1803–1863.* New Haven: Yale University Press.

Goldberg, D. E., and R. M. Turner. 1986. Vegetation Change and Plant Demography in Permanent Plots in the Sonoran Desert. *Ecology* 67:695–712.

Gray, J. P. 1940. When All Roads Led to Tombstone. Unpublished manuscript, Arizona Historical Society, Tucson.

Greeley, A. W., and W. A. Glassford. 1891. *Climate of Arizona with Particular Reference to Questions of Irrigation and Water Storage in the Arid Region.* H. ex. doc. 287, 51st Cong., 2nd sess. Washington, D.C.: GPO.

Greever, W. S. 1957. Railway Development in the Southwest. *New Mexico Historical Review* 32:151–203.

Griffin, J. R. 1971. Oak Regeneration in the Upper Carmel Valley, California. *Ecology* 52:862–868.

———. 1976. Regeneration in *Quercus lobata* Savannas, Santa Lucia Mountains, California. *American Midland Naturalist* 95:422–435.

———. 1977. Oak Woodland. In *Terrestrial Vegetation of California*, edited by M. G. Barbour and J. Major, 383–415. New York: John Wiley and Sons.

———. 1980. Sprouting in Fire-Damaged Valley Oaks, Chews Ridge, California. In *Proceedings of the Symposium on the Ecology, Management, and Utilization of California Oaks (June 26–28, 1979, Claremont, Calif.)*, coordinated by T. R. Plumb, 216–219. USDA Forest Service General Technical Report PSW-44. Berkeley, Calif.: Pacific Southwest Forest and Range Experiment Station.

Griffiths, D. A. 1901. *Range Improvement in Arizona.* USDA Bureau of Plant Industry Bulletin 4. Washington, D.C.: GPO.

———. 1910. *Range Improvement in Arizona.* USDA Bureau of Plant Industry Bulletin 177. Washington, D.C.: GPO.

Gritzner, J. A. 1988. *The West African Sahel: Human Agency and Environmental Change.* University of Chicago Geography Research Paper no. 226. Chicago: University of Chicago Press.

Haase, E. F. 1972. Survey of Floodplain Vegetation Along the Lower Gila River in Southwestern Arizona. *Journal of the Arizona Academy of Science* 7:66–81.

Hall, S. A. 1977. Late Quaternary Sedimentation and Paleoecologic History of Chaco Canyon, New Mexico. *Geological Society of America Bulletin* 88: 1593–1618.

Hamernick, D. M. and B. A. Brown. 1975. *Arizona's Remote Subdivisions: An Inventory.* Phoenix: Office of Economic Planning and Development, Office of the Governor, State of Arizona.

Harris, D. R. 1965. *Plants, Animals, and Man in the Outer Leeward Islands, West Indies: An Ecological Study of Antigua, Barbuda, and Anguilla.* University of California Publications in Geography, vol. 18. Berkeley: University of California Press.

———. 1966. Recent Plant Invasions in the Arid and Semiarid Southwest of the United States. *Annals of the Association of American Geographers* 56:408–422.

Harrison, A. E. 1972. The Santa Catalinas—A Description and History. Unpublished manuscript, Regional Office of the Coronado National Forest, Tucson.

Haskell, H. S. 1945. Successional Trends on a Conservatively Grazed Desert Grassland Range. *Journal of the American Society of Agronomists* 37:978–990.

Haskett, B. 1935. Early History of the Cattle Industry in Arizona. *Arizona Historical Review* 6:3–42.
———. 1936. History of the Sheep Industry in Arizona. *Arizona Historical Review* 7:3–49.
Hastings, J. R. 1959. Vegetation Change and Arroyo Cutting in Southeastern Arizona. *Journal of the Arizona Academy of Science* 1:60–67.
———. 1963. Historical Changes in the Vegetation of a Desert Region. Ph.D. diss., University of Arizona.
Hastings, J. R., and R. M. Turner. 1965. *The Changing Mile: An Ecological Study of Vegetation Change with Time in the Lower Mile of an Arid and Semiarid Region*. Tucson: University of Arizona Press.
Haury, E. W. 1976. *The Hohokam: Desert Farmers and Craftsmen. Excavations at Snaketown, 1964–1965*. Tucson: University of Arizona Press.
Haury, E. W., E. Antevs, and J. F. Lance. 1953. Artifacts with Mammoth Remains, Naco, Arizona. *American Antiquity* 19:1–24.
Haury, E. W., E. B. Sayles, and W. W. Wasley. 1959. The Lehner Mammoth Site, Southeastern Arizona. *American Antiquity* 25:2–30.
Heady, H. F. 1977. Valley Grassland. In *Terrestrial Vegetation of California*, edited by M. G. Barbour and J. Major, 491–514. New York: John Wiley and Sons.
Hecht, M. E., and R. W. Reeves. 1981. *Atlas of Arizona*. Tucson: Office of Arid Lands Studies, University of Arizona.
Hendrickson, D. A., and W. L. Minckley. 1984. Ciénegas—Vanishing Climax Communities of the American Southwest. *Desert Plants* 6:131–175.
Hennessy, J. T., R. P. Gibbens, J. M. Tromble, and M. Cardenas. 1983. Vegetation Changes from 1935 to 1980 in Mesquite Dunelands and Former Grasslands of Southern New Mexico. *Journal of Range Management* 36:370–374.
Herbel, C. H. 1985. Vegetation Changes on Arid Rangelands of the Southwest. *Rangelands* 7:19–21.
Hills, T. L. 1969. *The Savanna Landscapes of the Amazon Basin*. McGill University Savanna Research Project, Savanna Research Series, no. 14. Montreal: McGill University, Department of Geography.
Hinton, R. J. 1878. *The Handbook to Arizona: Its Resources, History, Towns, Mines, Ruins, and Scenery*. New York: American News Co.
Hoffmeister, D. F., and W. W. Goodpaster. 1954. *The Mammals of the Huachuca Mountains, Southeastern Arizona*. Illinois Biological Monographs, vol. 24. Urbana: University of Illinois Press.
Horton, J. S. 1964. *Notes on the Introduction of Deciduous Tamarisk*, USDA Forest Service Research Note RM-16. Fort Collins, Colo.: Rocky Mountain Forest and Range Experiment Station.
———. 1977. The Development and Perpetuation of the Permanent Tamarisk Type in the Phreatophyte Zone of the Southwest. In *Importance, Preservation, and Management of Riparian Habitat: A Symposium*, coordinated by R. R. Johnson and D. A. Jones, 124–127. USDA Forest Service General Technical Report RM-43. Fort Collins, Colo.: Rocky Mountain Forest and Range Experiment Station.
Horton, J. S., F. C. Mounts, and J. M. Kraft. 1960. *Seed Germination and Seed-*

ling Establishment of Phreatophyte Species. USDA Forest Service Paper 48. Fort Collins, Colo.: Rocky Mountain Forest and Range Experiment Station.

Houck, B., and S. Ambrose. 1978. Fuelwood Inventory and Management Plan. Unpublished report, Coronado National Forest, Douglas, Ariz.

Howe, H. C. n.d. Map of Cochise County—Arizona Territory. Arizona Historical Society, Tucson.

Humphrey, R. R. 1937. Ecology of Burroweed. *Ecology* 18:1–9.

——. 1949. Fire as a Means of Controlling Velvet Mesquite, Burroweed, and Cholla on Southern Arizona Ranges. *Journal of Range Management* 2:173–182.

——. 1953. The Desert Grassland, Past and Present. *Journal of Range Management* 6:159–164.

——. 1956. History of Vegetational Changes in Arizona. *Arizona Cattlelog* 11:32–35.

——. 1958. The Desert Grassland: A History of Vegetational Change and an Analysis of Causes. *Botanical Review* 24:193–252.

——. 1959. Lehmann's Lovegrass, Pros and Cons. In *Your Range—Its Management*, compiled by R. R. Humphrey, 28. University of Arizona Agricultural Experiment Station, Special Report no. 2. Tucson.

——. 1960. *Arizona Range Grasses: Their Description, Forage Value and Management.* Tucson: University of Arizona Press.

——. 1962. Fire as a Factor. In *Range Ecology*, edited by R. R. Humphrey, 148–189. New York: Ronald Press.

——. 1963. The Role of Fire in the Desert and Desert Grassland Areas of Arizona. *Proceedings—Tall Timbers Fire Ecology Conference*, no. 2, 45–62. Tallahassee, Fla.: Tall Timbers Research Station.

——. 1974. Fire in the Deserts and Desert Grassland of North America. In *Fire and Ecosystems*, edited by T. T. Kozlowski and C. E. Ahlgren, 365–400. New York: Academic Press.

——. 1987. *90 Years and 535 Miles: Vegetation Changes Along the Mexican Border.* Albuquerque: University of New Mexico Press.

Humphrey, R. R., and A. C. Everson. 1951. Effect of Fire on a Mixed Grass-Shrub Range in Southern Arizona. *Journal of Range Management* 4:264–266.

Humphrey, R. R., and L. A. Mehrhoff. 1958. Vegetation Change on a Southern Arizona Grassland Range. *Ecology* 39:720–726.

Huntington, E. 1914. *The Climatic Factor as Illustrated in Arid America.* Carnegie Institution of Washington, Publication 192. Washington, D.C.: Carnegie Institution.

International Boundary Commission (United States and Mexico, 1882–1896). 1898. *Views of the Monuments and Characteristic Scenes Along the Boundary Between the United States and Mexico West of the Rio Grande, 1892 to 1895.* Album in *Report of the Boundary Commission Upon the Survey and Remarking of the Boundary Between the United States and Mexico West of the Rio Grande, 1891 to 1896.* S. doc. 247, 55th Cong., 2nd sess. Washington, D.C.: GPO.

Johannessen, C. L. 1963. *Savannas of Interior Honduras.* Ibero-Americana, vol. 46. Berkeley: University of California Press.

Johnsen, T. N., and R. S. Dalen. 1984. Controlling Individual Junipers and Oaks with Pelleted Pialoran. *Journal of Range Management* 37:380–384.

Johnsen, T. N., and J. W. Elson. 1979. Sixty Years of Change on a Central Arizona Grassland-Juniper Woodland Ecotone. USDA Agricultural Research Service, *Agricultural Reviews and Manuals* ARM-W-7. Oakland, Calif.: Agricultural Research, Science and Education Administration, USDA.

Johnson, R. R., and D. A. Jones, eds. 1977. *Importance, Preservation, and Management of Riparian Habitat: A Symposium.* USDA Forest Service General Technical Report RM-43. Fort Collins, Colo.: Rocky Mountain Forest and Range Experiment Station.

Johnson, R. R., C. D. Ziebell, D. R. Patton, P. F. Ffolliott, and R. H. Hamre. 1985. *Riparian Ecosystems and Their Management: Reconciling Conflicting Uses.* USDA Forest Service General Technical Report RM-120. Fort Collins, Colo.: Rocky Mountain Forest and Range Experiment Station.

Johnston, M. C. 1963. Past and Present Grasslands of Southern Texas and Northeastern Mexico. *Ecology* 44:456–466.

Jordan, G. L., and M. L. Maynard. 1970. The San Simon Watershed—Historical Review. *Progressive Agriculture in Arizona* 22:10–13.

Judd, B. I., J. M. Laughlin, H. R. Guenther, and R. Handegarde. 1971. The Lethal Decline of Mesquite on the Casa Grande National Monument. *Great Basin Naturalist* 31:153–159.

Judd, N. M. 1931. Arizona's Prehistoric Canals from the Air. In *Smithsonian Explorations and Field-work in 1930.* Washington, D.C.: Smithsonian Institution.

Judson, S. 1952. Arroyos. *Scientific American* 187:71–76.

Karpiscak, M. M. 1980. Secondary Succession of Abandoned Field Vegetation in Southern Arizona. Ph.D. diss., University of Arizona.

Karpiscak, M. M., and O. M. Grosz. 1979. Dissemination Trails of Russian Thistle (*Salsola kali*) in Recently Fallowed Fields. *Journal of the Arizona-Nevada Academy of Science* 14:50–52.

Kates, R. W. 1987. The Human Environment: The Road Not Taken, the Road Still Beckoning. *Annals of the Association of American Geographers* 77:525–534.

Kearney, T. H., and R. H. Peebles. 1942. *Flowering Plants and Ferns of Arizona.* USDA Miscellaneous Publications no. 423. Washington, D.C.: GPO.

Kellogg, R. S. 1902a. Forest Conditions in Southern Arizona. *Forestry and Irrigation* 8:501–505.

———. 1902b. Report of an Examination of the Chiricahua Mountains in Arizona. Unpublished manuscript, Special Collections Library, University of Arizona.

———. 1902c. Report of an Examination of the Graham Mountains in Arizona. Open file report. Safford Range District, Coronado National Forest, Safford, Ariz.

Kelly, R. D., and B. H. Walker. 1976. The Effects of Different Forms of Land Use on the Ecology of a Semi-arid Region of Southeastern Rhodesia. *Journal of Ecology* 64:553–576.

Kennedy, C. B., M. M. Karpiscak, and M. C. Parton. 1986. Geographic Analysis of Agricultural Land Retirement, Cochise County, Arizona. In *Proceedings of the Conference on Remote Sensing and Geographic Information Systems in Management* (November 6–7, 1986), 4–6. Arizona Remote Sensing Center, Office of Arid Lands Studies, College of Agriculture, University of Arizona. Tucson.

Kline, V., and G. Cottam. 1979. Vegetation Response to Climate and Fire in the Driftless Area of Wisconsin. *Ecology* 60:861–868.

Komarek, E. V. 1968. The Nature of Lightning Fires. In *Proceedings—Tall Timbers Fire Ecology Conference*, no. 7, 5–41. Tallahassee, Fla.: Tall Timbers Research Station.

Lamb, W. A. n.d. Good Old Days in Tombstone, Arizona. Unpublished manuscript, Arizona Historical Society, Tucson.

Langton, J. 1904. The Power Plant of the Moctezuma Copper Company, Nacozari, Sonora, Mexico. *Transactions of the American Institute of Mining Engineers* 34:748–776.

Lanner, R. M. 1977. The Eradication of Pinyon-Juniper Woodland: Has the Program a Legitimate Purpose? *Western Wildlands* 4:12–17.

Lauver, M. E. 1938. A History of the Use and Management of the Forested Lands of Arizona 1862–1936. Master's thesis, University of Arizona.

Lehman, V. M. 1969. *Forgotten Legions: Sheep in the Rio Grande Plain of Texas.* El Paso: Texas Western University Press.

Lehr, J. H. 1978. *A Catalogue of the Flora of Arizona.* Phoenix: Desert Botanical Garden.

Leopold, A. 1921. A Plea for Recognition of Artificial Works in Forest Erosion and Control Policy. *Journal of Forestry* 19:267–273.

———. 1924. Grass, Brush, Timber, and Fire in Southern Arizona. *Journal of Forestry* 22:1–10.

Leopold, L. B. 1951a. Rainfall Frequency: An Aspect of Climatic Variation. *Transactions of the American Geophysical Union* 32:347–357.

———. 1951b. Vegetation of Southwestern Watersheds in the Nineteenth Century. *Geographical Review* 41:295–316.

Longhurst, W. M. 1956. Stump Sprouting of Oaks in Response to Seasonal Cutting. *Journal of Range Management* 9:194–196.

Longhurst, W. M., G. E. Connolly, B. W. Browning, and E. O. Garton. 1979. Food Interrelationships of Deer and Sheep in Parts of Mendocino and Lake Counties, California. *Hilgardia* 47:191–247.

Lowe, C. H. 1964. *Arizona's Natural Enrivonment: Landscapes and Habitats.* Tucson: University of Arizona Press.

McCarty, K. 1976. *Desert Documentary: The Spanish Years, 1776–1821.* Historical Monograph no. 4. Tucson: Arizona Historical Society.

McClintock, J. H. 1916. *Arizona: Prehistoric, Aboriginal, Pioneer, Modern.* 3 vols. Chicago: S. J. Clarke.

———. 1921. *Mormon Settlement in Arizona.* Phoenix: Manufacturing Stationers.

McDonald, J. E. 1956. *Variability of Precipitation in an Arid Region: A Survey of Characteristics for Arizona.* Technical Reports on the Meteorology and Climatology of Arid Regions, no. 1. Tucson: Institute of Atmospheric Physics, University of Arizona.

McGinnies, W. G. 1981. *Discovering the Desert*. Tucson: University of Arizona Press.

McLaughlin, S. P., and J. E. Bowers. 1982. Effects of Wildfire on a Sonoran Desert Plant Community. *Ecology* 63:246–248.

McQueen, I. S., and R. F. Miller. 1972. Soil-Moisture and Energy Relationships Associated with Riparian Vegetation near San Carlos, Arizona. *U.S. Geological Survey Professional Paper* 655-E:1–51.

Madany, M., and N. West. 1983. Livestock Grazing-Fire Regime Interactions Within Montane Forests of Zion National Park, Utah. *Ecology* 64:661–667.

Malin, J. C. 1953. Soil, Animal, and Plant Relations of the Grassland, Historically Reconsidered. *Scientific Monthly* 76:207–220.

Manje, J. M. 1954. *Luz de tierra incógnita (Unknown Arizona and Sonora 1693–1721)*. Translated by H. J. Karns and Associates. Tucson: Arizona Silhouettes.

Marsh, G. P. 1864. *Man and Nature; or, Physical Geography as Modified by Human Action*. New York: Charles Scribner.

———. 1874. *The Earth as Modified by Human Action*. New York: Scribner, Armstrong and Co.

Martin, P. S. 1967. Prehistoric Overkill. In *Pleistocene Extinctions, the Search for a Cause*, edited by P. S. Martin and H. E. Wright, Jr., 75–120. New Haven: Yale University Press.

Martin, S. C. 1975. *Ecology and Management of Southwestern Semi-Desert Grass-Shrub Ranges: The Status of Our Knowledge*. USDA Forest Service Research Papers RM-156. Fort Collins, Colo.: Rocky Mountain Forest and Range Experiment Station.

———. 1983. Responses of Semiarid Grasses and Shrubs to Fall Burning. *Journal of Range Management* 36:604–610.

Martin, S. C., and D. R. Cable. 1974. *Managing Semidesert Grass-Shrub Ranges: Vegetation Responses to Precipitation, Grazing, Soil Texture, and Mesquite Control*. USDA Forest Service Technical Bulletin 1480. Washington, D.C.: GPO.

Martin, S. C., and R. M. Turner. 1977. Vegetation Change in the Sonoran Desert Region, Arizona and Sonora. *Journal of the Arizona Academy of Science* 12:59–69.

Martineau, J. H. 1885. Settlements in Arizona. Unpublished manuscript, Bancroft Library, University of California, Berkeley.

Masse, W. B. 1981. Prehistoric Irrigation Systems in the Salt River Valley, Arizona. *Science* 214:408–415.

Matheny, R. L. 1975. The History of Lumbering in Arizona Before World War II. Ph.D. diss., University of Arizona.

Matlock, W. G., and P. R. Davis. 1972. *Groundwater in the Santa Cruz Valley Arizona*. University of Arizona Agricultural Experiment Station Technical Bulletin 194. Tucson: University of Arizona Press.

Mattison, R. H. 1946. Early Spanish and Mexican Settlements in Arizona. *New Mexico Historical Review* 21:273–327.

Meitl, J. M., P. L. Hathaway, and F. Gregg. 1983. *Alternative Uses of Arizona Lands Retired from Irrigated Agriculture*. Tucson: Cooperative Extension Service, College of Agriculture, University of Arizona.

México, Secretaría de Industria y Comercio, Dirección General de Estadística. 1963. *VIII Censo general de población, 1960: Estado de Sonora.* México, D. F.
———. 1971. *IX Censo general de población, 1970: Estado de Sonora.* México, D. F.
———. 1983. *X Censo general de población y vivienda 1980: Estado de Sonora.* Vol. 1. México, D. F.
Meyers, F. D. 1911. Cochise County. Unpublished manuscript, Arizona Historical Society, Tucson.
Miller, C. F. 1929. Prehistoric Irrigation Systems in Arizona. Master's thesis, University of Arizona.
Miller, R. R. 1961. Man and the Changing Fish Fauna of the American Southwest. *Papers of the Michigan Academy of Science, Arts, and Letters* 66:365–404.
Miller, W. N. n.d. Reminiscences. Unpublished manuscript, Arizona Historical Society, Tucson.
Minckley, W. L., and D. E. Brown. 1982. Wetlands. *Desert Plants* 4:222–287.
Minckley, W. L., and T. O. Clark. 1984. Formation and Destruction of a Gila River Mesquite Bosque Community. *Desert Plants* 6:23–30.
Moorhead, M. 1968. *The Apache Frontier: Jacobo Ugarte and Spanish–Indian Relations in Northern New Spain, 1769–1791.* Norman: University of Oklahoma Press.
Morrisey, R. J. 1950. The Early Range Cattle Industry in Arizona. *Agricultural History* 24:151–156.
Mowry, S. 1864. *Arizona and Sonora: The Geography, History, and Resources of the Silver Region of North America.* New York: Harper Brothers.
Myrick, D. F. 1975. *Railroads of Arizona.* Vol. 1, *The Southern Roads.* Berkeley: Howell-North Books.
Naveh, Z. 1967. Mediterranean Ecosystems and Vegetation Types in California and Israel. *Ecology* 48:443–459.
Neilson, R. P. 1986. High Resolution Climatic Analysis and Southwest Biogeography. *Science* 232:27–34.
Nelson, E. W. 1934. *The Influence of Precipitation and Grazing upon Black Grama Grass Range.* USDA Technical Bulletin 409. Washington, D. C.: GPO.
Nentvig, J. 1980. *Rudo Ensayo: A Description of Sonora and Arizona in 1764.* Translated, clarified, and annotated by A. F. Pradeau and R. R. Rasmussen. Tucson: University of Arizona Press.
Officer, J. E. 1987. *Hispanic Arizona, 1536–1856.* Tucson: University of Arizona Press.
Ohmart, R. D., and B. W. Anderson. 1982. North American Desert Riparian Ecosystems. In *Reference Handbook on the Deserts of North America,* edited by G. L. Bender, 433–479. Westport, Conn.: Greenwood Press.
Olmstead, F. H. 1919. *Gila River Flood Control—A Report on Flood Control of the Gila River in Graham County, Arizona.* S. doc. 436, 65th Cong., 3rd sess. Washington, D.C.: GPO.
Parker, K. W., and S. C. Martin. 1952. *The Mesquite Problem on Southern Arizona Ranges.* USDA Circular 908. Washington, D.C.: GPO.
Parker, M. B. 1979. *Mules, Mines and Me: 1895–1932.* Tucson: University of Arizona Press.

Parsons, J. J. 1955. The Miskito Pine Savanna of Nicaragua and Honduras. *Annals of the Association of America Geographers* 45:36–63.

———. 1981. Human Influences on the Pine and Laurel Forests of the Canary Islands. *Geographical Review* 71:253–269.

Pase, C. P., and C. E. Granfelt, technical coordinators. 1977. *The Use of Fire on Arizona Rangelands.* Arizona Interagency Range Committee Publication no. 4. Phoenix.

Patrick, H. R. 1903. *The Ancient Canal Systems and Pueblos of the Salt River Valley, Arizona.* Phoenix Free Museum Bulletin 1. Phoenix.

Paulsen, H. A. 1956. The Effect of Climate and Grazing on Black Grama. In *Ranch Day,* 17–24. Las Cruces, NM: New Mexico Agricultural Experiment Station and USDA Agricultural Research Service and Forestry Service.

Peters, E. D. 1911. *The Practice of Copper Smelting.* New York: McGraw-Hill.

Peterson, H. V. 1950. The Problem of Gullying in Western Valleys. In *Applied Sedimentation,* edited by P. D. Trask, 407–434. New York: John Wiley and Sons.

Phillips, F. J. 1912. *Emory Oak in Southern Arizona.* USDA Forest Service Circular 201. Washington, D.C.: GPO.

Phillips, W. S. 1963. Depth of Roots in Soil. *Ecology* 44:424.

Pinal Drill (Pinal City), Sept. 25, 1880.

Plumb, T. R. 1980. Response of Oaks to Fire. In *Proceedings of the Symposium on the Ecology, Management, and Utilization of California Oaks (June 26–28, 1979, Claremont, Calif.),* coordinated by T. R. Plumb, 202–215. USDA Forest Service General Technical Report PSW-44. Berkeley, Calif.: Pacific Southwest Forest and Range Experiment Station.

Potter, A. F. 1902. Report of the Examination of the Proposed Santa Rita Forest Reserve. Range Conditions in Arizona, 1900–1909, as Recorded by Various Observers in a Series of Miscellaneous Papers. Special Collections Library, University of Arizona, Tucson.

Pumpelly, R. 1918. *My Reminiscences.* 2 vols. New York: Holt and Co.

Pyne, S. J. 1982. *Fire in America.* Princeton: Princeton University Press.

———. 1984. *Introduction to Wildland Fire.* New York: John Wiley and Sons.

Range Conditions in Arizona, 1900–1909, as Recorded by Various Observers in a Series of Miscellaneous Papers. Special Collections Library, University of Arizona, Tucson.

Rea, A. M. 1983. *Once a River: Bird Life and Habitat Changes on the Middle Gila.* Tucson: University of Arizona Press.

Reynolds, H. G., and J. W. Bohning. 1956. Effects of Burning on a Desert Grass-Shrub Range in Southern Arizona. *Ecology* 37:769–777.

Reynolds, H. G., and G. E. Glendening. 1949. Merriam Kangaroo Rat a Factor in Mesquite Propagation on Southern Arizona Range Lands. *Journal of Range Management* 2:193–197.

Richardson, H. L. 1945. Discussion: The Significance of Terraces Due to Climatic Oscillation. *Geological Magazine* 82:16–18.

Robert, H. M. 1869. Map of Southern Arizona. Bancroft Library, University of California, Berkeley.

Robinson, T. W. 1965. Introduction, Spread, and Areal Extent of Saltcedar (*Tamarix*) in the Western States. *U.S. Geological Survey Professional Paper* 491-A:1–12.

Rockfellow, J. A. 1955. *Log of an Arizona Trail Blazer.* Tucson: Arizona Silhouettes.

Rodgers, W. M. 1965. Historical Land Occupance of the Upper San Pedro River Valley Since 1870. Master's thesis, University of Arizona.

Rogers, G. F. 1986. Comparison of Fire Occurrence in Desert and Nondesert Vegetation in Tonto National Forest, Arizona. *Madroño* 33:278–283.

Rogers, G. F., H. E. Malde, and R.M. Turner. 1984. *Bibliography of Repeat Photography for Evaluating Landscape Change.* Salt Lake City: University of Utah Press.

Rogers, G. F., and J. Steele. 1980. Sonoran Desert Fire Ecology. In *Proceedings of the Fire History Workshop (October 20–24, 1980) in Tucson, Arizona,* 15–19. USDA Forest Service General Technical Report RM-81. Fort Collins, Colo.: Rocky Mountain Forest and Range Experiment Station.

Rogers, G. F., and M. K. Vint. 1987. Winter Precipitation and Fire in the Sonoran Desert. *Journal of Arid Environments* 13:47–52.

Roskruge Collection. Photographs. Arizona Historical Society, Tucson.

Runnell, R. 1951. Some Effects of Livestock Grazing on Ponderosa Pine Forest and Range in Central Washington. *Ecology* 32:594–607.

Russell, R. P. 1982. The History of Man's Influence Upon the Vegetation of the Chiricahua Mountain Meadows. Master's thesis, University of Arizona.

Samuels, M. L., and J. L. Betancourt. 1982. Modeling the Long-term Effects of Fuelwood Harvests on Pinyon-Juniper Woodlands. *Environmental Management* 6:505–515.

Sauer, Carl O. 1932. *The Road to Cíbola.* Ibero-Americana, vol. 3. Berkeley: University of California Press.

———. 1935. *Aboriginal Population of Northwestern Mexico.* Ibero-Americana, vol. 10. Berkeley: University of California Press.

———. 1950. Grassland Climax, Fire, and Man. *Journal of Range Management* 3:16–21.

———. 1956. The Agency of Man on the Earth. In *Man's Role in Changing the Face of the Earth,* edited by W. L. Thomas, 49–69. Chicago: University of Chicago Press.

———. 1958. Man in the Ecology of Tropical America. *Proceedings of the Ninth Pacific Science Congress* 20:104–110. Bangkok, Thailand: Pacific Science Association.

Sawyer, D. A., and T. B. Kinraide. 1980. The Forest Vegetation at Higher Altitudes in the Chiricahua Mountains, Arizona. *American Midland Naturalist* 104:224–241.

Schmutz, E. M., B. N. Freeman, and R. E. Reed. 1968. *Livestock-Poisoning Plants of Arizona.* Tucson: University of Arizona Press.

Schroeder, M. J., and C. C. Buck. 1970. *Fire Weather: A Guide for Application of Meteorological Information to Forest Fire Control Operations.* U. S. Forest Service, Agriculture Handbook 360. Washington, D.C.: GPO.

Schwennesen, A. T. 1919. Groundwater in the San Simon Valley, Arizona and New Mexico. *U.S. Geological Survey Water Supply Paper* 425-A:1–28.

Scott, G. S. 1914. Bowie and San Simon. *Arizona: The New State Magazine* (Phoenix) 4:7.

Sellers, W. D. 1960. Precipitation Trends in Arizona and Western New Mexico.

Proceedings of the 28th Annual Western Snow Conference, 81–94. Fort Collins, Colo.: U.S. Soil Conservation Service.
Sellers, W. D., and R. H. Hill, eds. 1974. *Arizona Climate 1931–1972*. Tucson: University of Arizona Press.
Serven, J. E. 1965. *The Military Posts on Sonoita Creek*. The Smoke Signal, no. 12. Tucson: Tucson Corral of the Westerners.
Sheridan, T. E. 1986. *Los Tucsonenses: The Mexican Community in Tucson, 1854–1941*. Tucson: University of Arizona Press.
Shreve, F. 1942. The Vegetation of Arizona. In *Flowering Plants and Ferns of Arizona*, edited by T. H. Kearney and R. H. Peebles, 10–23. USDA Miscellaneous Publications, no. 423. Washington, D.C.: GPO.
———. 1951. *Vegetation and Flora of the Sonoran Desert*. Vol. 1, *Vegetation*. Carnegie Institution of Washington, Publication 591. Washington, D.C.: Carnegie Institution.
Skinner, W. W. 1903. *The Underground Waters of Arizona—Their Character and Uses*. University of Arizona Agricultural Experiment Station Bulletin 46.
Smith, G. E. P. 1910. *Ground Water Supply and Irrigation in the Rillito Valley*. University of Arizona Agricultural Experiment Station Bulletin 64.
Smith, N. J. H. 1981. Colonization Lessons from a Tropical Forest. *Science* 214:755–761.
Smith, W., and W. L. Heckler. 1955. Compilation of Flood Data in Arizona, 1862–1953. U.S. Geological Survey open file report. Tucson.
Sokal, R. R., and F. J. Rohlf. 1987. *Introduction to Biostatistics*. 2nd ed. San Francisco: W. H. Freeman.
Southwestern Stockman (Willcox), Sept. 14, 1889.
Spude, R. L. 1979. *Tombstone—Arizona Silver Camp*. Las Vegas: Nevada Publications.
Staunton, W. F. n.d. Mining papers (Box 5), The Tombstone Mill and Mining Co. (reports, 1881–1895). Special Collections, University of Arizona Library, Tucson.
Stewart, O. C. 1951. Burning and Natural Vegetation in the United States. *Geographical Review* 41:317–320.
———. 1956. Fire as the First Great Force Employed by Man. In *Man's Role in Changing the Face of the Earth*, edited by W. L. Thomas, 115–133. Chicago: University of Chicago Press.
Stoiber, P. E. 1973. Use of the U.S. General Land Office Survey Notes for Investigating Vegetation Change in Southern Arizona. Master's thesis, University of Arizona.
Stott, P. 1984. History of Biogeography. In *Biogeography*, edited by J. A. Taylor, 1–24. Totowa, N. J.: Barnes and Noble.
Surveyors' Field Notes, books 763, 779, 855, 882, 889, 918, 919, 942, 943, 956, 958, 1001, 1509, 1527, 1533, 1857, 2058, 2279, 3005, and 4024. U.S. Bureau of Land Management, Phoenix.
Swetnam, T. W. (In press). Fire History and Climate in the Southwestern United States. In *Effects of Fire in Management of Southwestern Natural Resources*. USDA Forest Service General Technical Report. Fort Collins, Colo.: Rocky Mountain Forest and Range Experiment Station.

Bibliography

Swetnam, T. W., C. H. Baisan, P. M. Brown, and A. C. Caprio. 1989. Fire History of Rhyolite Canyon, Chiricahua National Monument. Unpublished manuscript prepared for the National Park Service, Southern Arizona Group Office (contract PX 8601-7-0106), Phoenix.

Tainter, F. H., S. W. Fraedrich, and D. M. Benson. 1984. The Effect of Climate on Growth, Decline, and Death of Northern Red Oaks in the Western North Carolina Nantahala Mountains. *Castanea* 49:127–137.

Tenney, J. B. n.d. Tombstone Mining History. Unpublished manuscript, Arizona Historical Society, Tucson.

Thomas, W. L., ed. 1956. *Man's Role in Changing the Face of the Earth.* Chicago: University of Chicago Press.

Thornber, J. J. 1906. Alfilaria (*Erodium cicutarium*), as a Forage Plant in Arizona. *University of Arizona Agricultural Experiment Station Bulletin* 52:27–58.

———. 1907. *18th Annual Report.* Tucson: University of Arizona Experiment Station.

———. 1910. The Grazing Ranges of Arizona. *University of Arizona Agricultural Experiment Station Bulletin* 65:245–360.

Thornthwaite, C. W., C. F. Sharpe, and E. F. Dosch. 1942. *Climate and Accelerated Erosion in the Arid and Semiarid Southwest, with Special Reference to the Polacca Wash Drainage Basin, Arizona.* USDA Technical Bulletin 808. Washington, D.C.: GPO.

Tombstone Daily Nugget, June 10, 1880; Dec. 31, 1881; Feb. 5, 1882; Feb. 18, 1882; May 9, 1882; May 21, 1882.

Tombstone Epitaph, Aug. 27, 1880; Sept. 17, 1881; Apr. 3, 1882.

Tombstone Prospector, Oct. 19, 1889.

Tombstone Weekly Epitaph, June 12, 1880.

Toumey, J. W. 1891a. *I. Notes of Some of the Range Grasses of Arizona.* University of Arizona Agricultural Experiment Station Bulletin 2.

———. 1891b. *II. Overstocking the Range.* University of Arizona Agricultural Experiment Station Bulletin 2.

———. 1901. Our Forest Reservations. *Popular Science Monthly* 59:125–126.

Towne, D. C. 1986. The Relationship Between Irrigated Farmland Decline and Physical Landscape Factors: A Spatial Analysis. Master's thesis, University of Arizona.

Tschirley, F. H., and S. C. Martin. 1961. *Burroweed in Southern Arizona Rangelands.* Arizona Agricultural Experiment Station Technical Bulletin 146.

Tucson Daily Record, May 13, 1880; May 15, 1880; June 2, 1880; June 4, 1880.

Turner, R. M. 1974. Quantitative and Historical Evidence of Vegetation Changes Along the Upper Gila River, Arizona. *U.S. Geological Survey Professional Paper* 655-H:1–20.

Turner, R. M., and J. E. Bowers. 1987. Long-term Changes in Populations of *Carnegiea gigantea*, Exotic Plant Species, and *Cercidium floridum* at the Desert Laboratory, Tumamoc Hill, Tucson, Arizona. In *Arid Lands: Today and Tomorrow, Proceedings of the International Arid Lands Research and Development Conference, Oct. 1985, Tucson.* Boulder, Colo.: Westview Press.

U.S. Bureau of Land Management, Department of the Interior. 1973. *Manual of Instructions for the Survey of the Public Lands of the United States.* USDI Technical Bulletin no. 6. Washington, D.C.: GPO.

———. 1978. *Final Environmental Statement—Upper Gila–San Simon.* Washington, D.C.: GPO.
U.S. Bureau of the Census. 1864. *Eighth Census of the United States, 1860: Population.* Washington, D.C.: GPO.
———. 1872. *Ninth Census of the United States, 1870: Population.* Washington, D.C.: GPO.
———. 1883a. Report on Cattle, Sheep, and Swine. *Tenth Census of the United States, 1880: Report on the Production of Agriculture.* Washington, D.C.: GPO.
———. 1883b. *Tenth Census of the United States, 1880: Population.* Washington, D.C.: GPO.
———. 1902. *Twelfth Census of the United States, 1900: Agriculture.* Vol. 6, pt. 2. Washington, D.C.: GPO.
———. 1913. *Thirteenth Census of the United States, 1910: Agriculture.* Vol. 6. Washington, D.C.: GPO.
———. 1922. *Fourteenth Census of the United States, 1920: Agriculture.* Vol. 6, pt. 3. Washington, D.C.: GPO.
———. 1943. *Sixteenth Census of the United States, 1940: Housing.* Vol. 2. Washington, D.C.: GPO.
———. 1975. *Historical Statistics of the United States, Colonial Times to 1970.* Pt. 2. Washington, D.C.: GPO.
———. 1983. *Twentieth Census of the United States, 1980.* Vol. 1, *Characteristics of the Population.* Chap. D, Detailed Population Characteristics, pt. 4, Arizona. Washington, D.C.: GPO.
U.S. Congress, House. 1879. *Annual Report of the Governor of Arizona.* H. ex. doc. 1, 46th Cong., 2nd sess. Serial 1911.
———. 1881. *Annual Report of the Governor of Arizona.* H. ex. doc. 1, 47th Cong., 1st sess. Serial 2018.
———. 1883. *Annual Report of the Governor of Arizona.* H. ex. doc. 1, 48th Cong., 1st sess. Serial 2191.
———. 1885. *Annual Report of the Governor of Arizona.* H. ex. doc. 1, 49th Cong., 1st sess. Serial 2379.
———. 1890. *Annual Report of the Governor of Arizona.* H. ex. doc. 1, 51st Cong., 2nd sess. Serial 2642.
———. 1893. *Annual Report of the Governor of Arizona.* H. ex. doc. 1, 53rd Cong., 2nd sess. Serial 3211.
———. 1895. *Annual Report of the Governor of Arizona.* H. ex. doc. 5, 54th Cong., 1st sess. Serial 3383.
U.S. Department of Agriculture and Arizona Water Commission. 1977. *Santa Cruz-San Pedro River Basin, Arizona: Resource Inventory.* Portland, Ore.
———. 1986. *Arizona Agriculture: Now a Vision of the Future.* Tucson: College of Agriculture, University of Arizona.
Upson, A., W. J. Cribbs, and E. B. Stanley. 1937. Occurrence of Shrubs on Range Areas in Southeastern Arizona. A Condensed Report of the Results of a Cooperative Survey by the Arizona Agricultural Experiment Station, Southwest Forest and Range Experiment Station, Agriculture Adjustment Administration and Bureau of Agricultural Economics. Issued by Arizona Experiment Station. Mar. 1953.

Vale, T. R. 1977. Forest Changes in the Warner Mountains, California. *Annals of the Association of American Geographers* 67:28–45.

———. 1981. Ages of Invasive Trees in Dana Meadows, Yosemite National Park, California. *Madroño* 28:45–47.

———. 1982. *Plants and People: Vegetation Change in North America*. Washington, D.C.: Association of American Geographers.

Van Sickle, C., and G. Newman. 1981. Fuelwood Inventory—Sierra Vista District. Report, Coronado National Forest, Sierra Vista, Ariz.

Van Vegten, J. A. 1983. Thornbush Invasion in a Savanna Ecosystem in Eastern Botswana. *Vegetatio* 56:3–7.

Veblen, T. T., and C. J. Steward. 1982. The Effects of Introduced Wild Animals on New Zealand Forests. *Annals of the Association of American Geographers* 72:372–397.

Vogl, R. J. 1971a. The Future of Our Forest. *Ecology Today* 1:6–8, 58.

———. 1971b. Smokey: A Bearfaced Lie. *Ecology Today* 1:14–17.

Voigt, W., Jr. 1976. *Public Grazing Lands: Use and Misuse by Industry and Government*. New Brunswick, N.J.: Rutgers University Press.

Wagoner, J. J. 1951. Development of the Cattle Industry in Southern Arizona, 1870s and 1880s. *New Mexico Historical Review* 26:204–224.

———. 1952. *History of the Cattle Industry in Southern Arizona, 1540–1940*. University of Arizona Social Science Bulletin 20. Tucson: University of Arizona Press.

———. 1961. Overstocking of the Ranges in Southern Arizona During the 1870s and 1880s. *Arizoniana* 2:23–27.

———. 1970. *Arizona Territory 1863–1912: A Political History*. Tucson: University of Arizona Press.

———. 1975. *Early Arizona*. Tucson: University of Arizona Press.

Walker, B. H., D. Ludwig, C. S. Holling, and R. M. Peterman. 1981. Stability of Semiarid Savanna Grazing Systems. *Journal of Ecology* 69:473–498.

Wallmo, O. C. 1955. Vegetation of the Huachuca Mountains, Arizona. *American Midland Naturalist* 54:466–480.

Warren, D. K., and R. M. Turner. 1975. Saltcedar (*Tamarix chinensis*) Seed Production, Seedling Establishment, and Response to Inundation. *Journal of the Arizona Academy of Science* 10:135–144.

Waters, M. R. 1988. Holocene Alluvial Geology and Geoarchaeology of the San Xavier Reach of the Santa Cruz River, Arizona. *Geological Society of America Bulletin* 100:479–491.

Weaver, H. 1947. Fire—Nature's Thinning Agent in Ponderosa Pine. *Journal of Forestry* 45:437–444.

———. 1951a. Fire as an Ecological Factor in the Southwestern Ponderosa Pine Forests. *Journal of Forestry* 49:93–98.

———. 1951b. Observed Effects of Burning on Perennial Grasses in the Ponderosa Pine Forests. *Journal of Forestry* 49:267–271.

———. 1955. Fire as an Enemy, Friend, and Tool in Forest Management. *Journal of Forestry* 53:499–504.

———. 1968. Fire and Its Relationship to Ponderosa Pine. In *Proceedings—Tall Timbers Fire Ecology Conference*, no. 7, 127–149. Tallahassee, Fla.: Tall Timbers Research Station.

---. 1974. Effects of Fire on Temperate Forests: Western United States. In *Fire and Ecosystems*, edited by T. T. Kozlowski and C. E. Ahlgren, 279–319. New York: Academic Press.
Weber, D. J. 1982. *The Mexican Frontier, 1821–1846: The American Southwest Under Mexico*. Albuquerque: University of New Mexico Press.
Weech, G., comp. 1979. *Pioneer Town: Pima Centennial History*. Pima, Ariz.: Eastern Arizona Museum and Historical Society.
Weeden, P. 1980. Fuelwood Inventory—Dragoon Mountains. Report, Coronado National Forest, Douglas, Ariz.
---. 1981. Fuelwood Inventory—Peloncillo Mountains. Report, Coronado National Forest, Douglas, Ariz.
Weekly Arizonan (Tucson), May 28, 1870.
Weekly Arizonian (Tubac), June 2, 1859; June 9, 1859.
Weekly Nugget (Tombstone), June 3, 1880; June 10, 1880.
Werger, M. J. A. 1980. A Phytosociological Study of the Upper Orange River Valley. *Memoirs of the Botanical Survey of South Africa* 46:1–92.
West, R. C. 1949. *The Mining Community in Northern New Spain: The Parral Mining District*. Ibero-Americana, vol. 30. Berkeley: University of California Press.
White, K. L. 1966. Structure and Composition of Foothill Woodland in Central Coastal California. *Ecology* 47:229–237.
White, L. D. 1969. Effects of a Wildfire on Several Desert Grassland Shrub Species. *Journal of Range Management* 22:284–285.
Whitfield, C. J., and H. L. Anderson. 1938. Secondary Succession in the Desert Plains Grassland. *Ecology* 19:171–180.
Whitfield, C. J., and E. L. Beutner. 1938. Natural Vegetation in the Desert Plains Grassland. *Ecology* 19:26–37.
Williams, C. P. 1964. Lehmann Lovegrass-Velvet Mesquite Invasion Relationships in the Desert Grassland. Master's thesis, University of Arizona.
Willson, R. 1966. Pioneers Chopped Their Hay. *Arizona Republic* (Phoenix), July 3.
Wilson, E. D. 1961. *Gold Placers and Placering in Arizona*. Arizona Bureau of Mines Bulletin 168. Tucson: University of Arizona.
Winn, F. J. n.d. Papers. Arizona Historical Society, Tucson.
Woodward, S. L. 1972. Spontaneous Vegetation of the Murray Springs Area, San Pedro Valley, Arizona. *Journal of the Arizona Academy of Science* 7:12–16.
Woodward, S. M. 1904. *Cost of Pumping for Irrigation*. University of Arizona Agricultural Experiment Station Bulletin 49.
Wooton, E. O. 1916. *Carrying Capacity of Grazing Ranges in Southern Arizona*. USDA Bulletin no. 367. Washington, D.C.: GPO.
Wright, H. A. 1980. *The Role and Use of Fire in the Semidesert Grass-Shrub Type*. USDA Forest Service General Technical Report INT-85. Ogden, Utah: Intermountain Forest and Range Experiment Station.
Wright, H. A., and A. W. Bailey. 1982. *Fire Ecology*. New York: John Wiley and Sons.
Wright, H. A., S. C. Bunting, and L. F. Neuenschwander. 1976. Effect of Fire on Honey Mesquite. *Journal of Range Management* 29:467–471.
Wynne, F. 1926. The West Fork of the Gila River. *Science* 64:16–17.

Young, J. A., and J. D. Budy. 1979. Historical Use of Nevada's Pinyon-Juniper Woodlands. *Journal of Forest History* 23:113–121.

York, J. C., and W. A. Dick-Peddie. 1969. Vegetation Changes in Southern New Mexico During the Past Hundred Years. In *Arid Lands in Perspective*, edited by W. G. McGinnies and B. J. Goldman, 157–166. Tucson: University of Arizona Press.

Zohary, M. 1973. *Geobotanical Foundation of the Middle East*. Stuttgart: Gustav Fischer Verlag.

Zwolinski, M. J., and J. H. Ehrenreich. 1968. Prescribed Burning on Arizona Watersheds. In *Proceedings—Tall Timbers Fire Ecology Conference*, no. 7, 195–205. Tallahassee, Fla.: Tall Timbers Research Station.

Index

References to illustrations are set in italics.

Abandoned fields, 39–40, 98, 99, 160, 164–65
Abies concolor, 27
Abies lasiocarpa, 27
Acacia, increase of, 24, 53, 63, 66, 69, 71, *81*, 87, 99, 119–20, 135, 174, 179, 187. *See also* Brush invasion; Catclaw; Viscid acacia
Acacia greggii. *See* Acacia; Catclaw
Acacia neovernicosa, 22. *See also* Acacia
Aerial photography, 59, 60, 90–100, 103, 176, 177, 187; NHAP, 90, 91, 96, 99
Agave spp., 24
Agricultural clearing. *See under* Agriculture
Agriculture: artesian wells, 37; dry farming, 39, 162; during Spanish/Mexican occupation, 31; early Anglo, 33–34, 37, 161–64; effects of agricultural clearing, 15, 39–44 *passim*, 66, 98, 161, 165, 172, 173, 177–78, 180, 184; modern irrigated, 39, 40, 43, 44, 46, 47, 109–11, 161, 162–66, 178; prehistoric Indian, 29, 30. *See also* Abandoned fields
Alfilaria. *See* Filaree
Allen Flat, 25
Alligator juniper, 25
Alto, 79

Alto Mine, 73
Amaranth, 24
Amaranthus palmeri, 165
Amaranthus spp., 24
Ambrosia deltoidea. *See* Bursage
Ambrosia spp. *See* Bursage
Andropogon spp., 25
Antelope, 4
Apache Indians, 30, 32–34, 59; arrival in southeastern Arizona, 31; pacification of, 34, 35, 116, 162; raids by, 30–34 *passim*, 114, 145, 186; setting of fires by, 125, 127–29; wild hay harvest by, 173–74
Apache Pass, 34
Apache pine, 25, 166
Aravaipa Canyon, *81*, 128
Aravaipa (Creek) Valley, 18, 20; livestock grazing in, 112; modern farming in, 37, 156–57, 161, 162; prehistoric agricultural settlement in, 30; vegetation change in, 179. *See also* Fort Breckenridge
Arbutus arizonica, 26
Arctostaphylos spp. *See* Manzanita
Aridity. *See under* Climate
Aristida spp. *See* Threeawn
Arizona Experiment Station, 109, 113, 157
Arizona sycamore, 28. *See also* Riparian wetlands
Arizona upland. *See* Sonoran desertscrub
Arizona white oak, 25
Arizona-Pittsburg Mine, 79

Arizona-Sonora boundary, 47, 48, 49, 67, 186. *See also* International boundary
Arroyo cutting, 15, 19, 33, 43, 46, 122–23, 176; causes of, 18–19, 44–48, 58, 111, 115, 117–18, 121, 161, 165, 185, 196
Artesian wells. *See under* Agriculture
Ash Canyon, 171
Atriplex canescens, 21
Avra Valley, 165

Babocomari (Creek) Valley, 34, 71; livestock grazing in, 115; modern farming in, 37, 162; prehistoric agricultural settlement in, 30; San Ignacio del Babocomari land grant, 114, 116; woodcutting in, 147
Barfoot (Park) Canyon, 171, 184
Barley, 155
Barrel Canyon, 83
Beaver, 4, 44, 46, 47, 111
Bellota Ranch, 61, 62
Benson, 36, 41, 144, 145, 151, 164
Bermuda grass, 155, 156
Bisbee, 36, 38, 41, 148, 151, 168, 170, 174, 184; Copper Queen Mine, 144, 145, 150; fuelwood consumption in, 143, 145
Black Diamond, 145
Black grama, 139. *See also* Grama
Black willow, 89. *See also* Riparian wetlands
Blue paloverde, 178
Bluestem, 25
Boerhaavia spp., 24
Bosques, 36, 47, 48, 157, 166, 176. *See* Mesquite
Boston, 36, 147
Bouteloua eripoda, 139
Bouteloua spp. *See* Grama
Bowie, 34, 162
Brittlebush, 22, 178
Bromus rubens. *See* Red brome
Broom snakeweed. *See* Snakeweed
Brush invasion, 3, 4, 42, 47, 50, 66–67; causes of, 50, 53, 56, 90, 98–100; control of, 99, 119, 122, 125; of desertscrub, 98–99, 142, 155, 156, 178–79; effects of fire on, 140; of evergreen woodland, 44, 180–82, 186; of grasslands, 22–24, 43–45, 48, 52–53, 56, 57, 62–63, 69, 81, 90, 99, 113–15, 119–20, 125, 129, 135, 138–39, 155, 173, 174, 178–80, 185, 186; repeat photography of, 52, 69, 71, 73, 75, 81, 87, 90, 98–99; of riparian wetland, 28, 39, 56, 155, 157–58, 176, 177. *See also* Acacia; Burroweed; Fire suppression; Livestock grazing; Mesquite; Snakeweed
Bryce, 37, 162
Buckbush, 26
Buckwheat, 24
Buffelgrass, 160
Bureau of Land Management, 118–19, 120, 196
Burroweed, 49, 63, 85; cyclic fluctuations in, 67; effect of fire on, 139–40; increase of, 4, 24, 53, 81, 179; spread of, by livestock, 119–20. *See also* Brush invasion
Bursage, 21, 22, 49, 67, 178
Bush muhly, 24

Calabasas, 32, 33, 34, 73, 115–16, 162
Calliandra eriophylla, 178
Camp Bowie, 34, 162
Camp Crittenden, 34, 77
Camp Grant. *See* Fort Grant
Camp Wallen, 34, 162
Cañada del Oro, 61
Cananea, Sonora, 151
Canelo Hills, 135
Carelessweed, 165
Carnegiea gigantea. *See* Sahuaro
Casa Grande, 116, 134, 162
Catclaw, 22, 61, 112. *See also* Acacia
Cattle: effects of 1891–93 drought on, 37, 109, 117, 186; effects on vegetation, 32–33, 43–44, 46–47, 75, 81, 85, 109, 111, 115, 119–21, 182, 185, 195; numbers of, after 1870, 37, 57, 112, 115–18, 183, 185–86; numbers of, before 1870,

14–15, 32, 33, 114–15, 185; role in erosion, 111, 113, 115, 118. *See also under* Arroyo cutting; Livestock grazing
Ceanothus greggii, 26
Ceanothus spp., 145
Cenchrus ciliaris, 160
Cercidium floridum, 178
Cercidium microphyllum. See Foothills paloverde
Cercocarpus montanus, 26, 145
Chaparral, 26, 179. *See also* Evergreen woodland
Charcoal, making of, 34, 46, 144, 169
Charleston, 36, 69, 128, 147
Chihuahua pine, 25
Chihuahuan desertscrub, 20, 22, 23, 52, 59, 61, 63, 91, 97–100, 141–42, 178–79. *See also* Desertscrub
Chilopsis linearis. See Desert willow
Chiricahua Mountains, 18, 26–27, 35; fires in, 138; forest reserve, 38, 182–83; livestock grazing in, 117; logging in, 37, 144, 147–48, 166, 170–72, 184; vegetation change in, 180, 184
Chiricahua Peak, 18
Chloris virgata, 160
Ciénaga. *See under* Riparian wetlands
Ciénega Creek, 156, 162
Ciénega Ranch, 117
Clearcutting. *See under* Logging
Cliff rose, 26
Clifton-Morenci, 36, 38, 144, 145, 148
Climate: change and fluctuations, 14–15, 18, 42, 43, 47–58 *passim*, 100–5, 139, 185; description of, 19–20, 195; trend toward aridity, 3, 15, 53, 100–5, 178, 180, 185, 186
Cochise County, 20, 40; fires in, 130; livestock grazing in, 37, 117; modern farming in, 119, 164; wild hay harvest in, 173–74; woodcutting in, 147–49
Colorado River, 50, 52, 104, 195
Colorado squawfish, 177

Conifer forest, mixed. *See* Mixed-conifer Forest
Contention (Mill), 36, 69, 146, 147
Coolidge, 39, 165
Coolidge Dam, 39, 67, 164
Corkbark fir, 27
Coronado National Forest, 38, 49, 66–67, 75, 79, 122, 124, 196; fires in, 126, 127, 128, 137; logging in, 147, 151, 152, 171, 180. *See also* Fire; Fire suppression
Cotton grass, 24
Cottonwood, 4, 28, 56, 66–67, 69, 71, 89, 111, 144, 153, 165, 176–77. *See also* Riparian wetlands
Courtland, 38, 145
Cowania mexicana. See Cliff rose
Creosote bush, 21, 22, 119, 139, 178
Crittenden, 35, 162
Croton Springs, 180
Crucifixion thorn, 22
Curly mesquite, 24. *See also* Mesquite
Cynodon spp. *See* Bermuda grass

Dasylirion spp., 24
Davis Ranch, 73
Desert broom, 120
Desert grasslands, 47–50 *passim*, 67, 90, 125, 138–40, 147, 172. *See also* Grasslands
Desert willow, 67, 144. *See also* Riparian wetlands
Desertscrub: changes in, 56, 98–100, 139, 161, 164, 178–79; description of, 20–23, 143, 147, 178–79; fire in, 129, 135–37, 141–42, 156, 178; repeat photography of, 47–52, 59, 91, 97–98. *See also* Brush invasion; Chihuahuan desertscrub; Sonoran desertscrub
Devil's claw, 24
Dos Cabezas, 145
Douglas, 41, 115, 151, 196
Douglas fir, 27, 138, 166
Downing Canyon, 171
Dragoon Mountains, 35; fires in, 131, 132, 135; forest reserve, 38;

vegetation change in, 174, 180; woodcutting in, 144–48, 152
Drought of 1891–93: effects on ranching, 37, 109, 117, 186; effects on vegetation, 3, 55, 77, 112, 120, 138, 139, 186
Dry farming. *See under* Agriculture
Duncan, 37, 115–16
Duncan Valley, 19
Duquesne, 145

El Paso, 35
Elgin, 49, 159
Emery City, 147
Emory oak, 25, 154
Empire Ranch, 63
Encelia farinosa, 22, 178
Engelmann spruce, 27
Eragrostis chloromelas, 160
Eragrostis curvula, 160
Eragrostis intermedia, 25
Eragrostis lehmanniana. *See* Lehmann lovegrass
Eriogonum spp., 24
Erodium circutarium. *See* Filaree
Erodium spp. *See* Filaree
Erosion, 33, 39, 117, 120, 121–23, 155–56, 160–61. *See also* Agriculture; Arroyo cutting; Livestock grazing
Evergreen woodland, 179; changes in, 44, 48–52, 56, 62–63, 66–67, 73, 99, 120, 123, 152–54, 180–82, 186–87; description of, 20, 25–26; effects of livestock grazing in, 113, 118–19, 182; fires in, 125–29, 135–37, 142, 181–82; Forest Service management of, 99, 119; repeat photography of, 47–49, 52, 56, 59, 73, 75, 77, 79, 83, 91; woodcutting in, 143–54, 180, 182, 196
Exotics, introduction and spread of, 4, 15, 24, 42, 49, 53, 90, 99, 108, 119, 122, 123, 141, 155–60 *passim*, 161, 165, 174, 176–80, 184, 186, 196

Fairbank, 36, 61, 71, 144
Fairy duster, 178

Filaree, 24, 112, 141, 155, 156–57, 160
Fingergrass, 174
Fire, 124–42; effects of, 104, 111, 122, 129–35, 140–41, 153, 183; frequency of, 4, 14, 37, 44, 124–29, 135–42, 156, 178–79; sources of, 28–30, 46, 122, 124–29, 133–35, 138–39, 181, 183. *See also* Fire suppression; Prescribed burning
Fire suppression, 15, 37, 43, 53, 90, 119, 128–29, 137–39; as a result of overgrazing, 44, 115, 120, 124–25, 128; effects of, on vegetation, 3, 42, 44, 46, 57, 66, 75, 99, 105, 120, 122, 125, 137–39, 142, 180–87 *passim*. *See also* Fire; Prescribed burning
Florence, 18, 21, 39, 41, 135, 142, 151, 162, 168, 178
Florence Canal, 162–63
Flourensia cernua, 22
Foothills paloverde, 22, 49, 178. *See also* Paloverde
Forest Reserves. *See* National Forest Reserves
Forest Service, 148; management of wildlands by, 39, 99, 118–19, 120–21, 153, 182–83
Fort Bowie, 34, 162
Fort Breckenridge, 34
Fort Buchanan, 34, 77
Fort Grant, 35, 162, 169, 173, 174
Fort Huachuca, 36
Fort Lowell, 102
Fouquieria splendens, 22
Fraxinus pennsylvanica, 28
Fraxinus spp., 28
Frye Canyon, 171
Fuelwood. *See* Woodcutting

Gadsden Purchase, 32, 34, 115
Galeria forest. *See* Riparian wetlands
Galeyville, 145
Galiuro Mountains, 26
Galleta grass, 21, 25, 174
Gambel oak, 27
Garden Canyon, 171
Gardner Canyon, 171

Garrya spp., 145
Garrya wrightii, 26
General Land Office. *See* Surveyors' field notes
Gila River Basin, 18, 20, 124, 176; Apache Indian settlement of, 31; early Anglo settlement of, 32, 35–37, 46–47, 162; livestock grazing in, 38, 115–16, 121–23; modern farming in, 27, 39, 162–65; Pima Indian settlement of, 29, 33; prehistoric agricultural sites in, 30; Sobaípuri Indian settlement of, 29, 31; tamarisk invasion of, 39, 67, 157–58
Gila River Phreatophyte Project, 158
Gleeson, 38
Globe, 36, 38, 145
Globe Mine (Old Dominion), 36, 144
Government Draw, 174
Graham County, 117
Graham Mountains. *See* Pinaleño (Graham) Mountains
Grama, 24–25, 111, 123, 139, 156, 174, 175
Grand Central, 147, 152
Grand View Peak, 171
Granite burn, 135, 141
Grasslands: changes in, 42–53, 69, 109–13, 119, 122, 139, 161, 164, 179–80, 185; description of, 4, 23–25, 37; effects of fire on, 125, 128–29, 135–42; repeat photography of, 50–52, 59, 69, 71, 91, 179. *See also* Brush invasion; Desert grasslands; Livestock grazing; Plains grasslands; Semidesert grasslands; Wild hay harvesting
Grazing. *See* Livestock grazing
Greaterville, 75, 145
Green Mountain, 171
Greenbush Draw, 179
Grizzly bear, 4
Ground photography, 4, 15, 42, 45, 55, 59, 176, 177, 179, 181–82, 185–87; methodological problems with, 5, 14, 16, 55, 67; studies using, 43–44, 47–56, 63–90. *See also* Aerial photography
Groundwater, overdrafts of, 39–40, 42, 47, 90, 158, 161, 163–66, 176–77
Guébavi, 31
Gutierrezia sarothrae. See Snakeweed

Haplopappus tenuisectus. See Burroweed
Harshaw, 36, 144, 145, 149, 168
Hay, wild. *See* Wild hay harvesting
Headcenter, 147
Heintzelman, 144
Heliograph Peak, 171
Helvetia, 36, 145
Hereford, 37, 145, 162
Hilaria belangeri, 24
Hilaria jamesii, 25
Hilaria mutica. See Tobosa
Hilaria rigida, 21
Hilltop, 145
Hog Canyon, 77
Hohokam Indians, 29
Honey mesquite, 22, 28. *See also* Mesquite
Hordeum vulgare, 155
Horses, 31, 32, 33, 46, 47, 111, 114, 115
Huachuca Mountains, 26, 174, 179–80; fires in, 128, 131, 133, 134; forest reserve, 38; logging in, 36–37, 147, 166–69, 171, 183, 184; woodcutting in, 144, 152
Huerfano Butte, 87

Indians: impact of, 29–33; setting of wildfires by, 28–30, 46, 124–29 *passim*, 133–35, 181. *See also* Apache Indians; Hohokam Indians; Pima Indians; Sobaípuri Indians; Tohono O'odham Indians
International boundary: comparison of climate across, 100; comparison of vegetation across, 32–33, 42, 49–52, 56, 89, 104, 114, 151, 154, 179, 181; monument markers of,

52, 89, 195. *See also* Arizona-Sonora boundary
Ironwood, 22, 178

Jacobsen's Mill, 171
Janusia gracilis, 178
Johnson, 145
Jornada Experimental Range, 59
Josephine Canyon, 79
Juglans major, 28
Juniper, 25, 62; control of, 119, 125; effects of woodcutting on, 144–45, 148, 153, 154, 170, 182; spread of, 24, 44, 56, 66–67, 83, 119–21, 125, 179, 186. *See also* Evergeen woodland; One-seed juniper
Juniperus deppeana, 25
Juniperus monosperma. *See* One-seed juniper

Kansas Settlement, 162
Koeberlinia spinosa, 22
Krameria grayi, 178

Land grants, Mexican and Spanish, 32–33, 69, 71, 73, 89, 114, 116, 117, 186. *See also individual grants*
Larrea tridentata. *See* Creosote bush
Lehmann lovegrass, 24, 53, 99, 155–60 *passim*
Lightning fires. *See* Fire, sources of
Littleleaf sumac, 22. *See also* Sumac
Live oak, 137
Livestock grazing: control of, 38–39, 45, 99, 118–19, 122–23, 153, 156, 160, 180, 182–83; during early Anglo settlement, 35–36, 114–17, 179–80; during Spanish and Mexican occupations, 30–33, 46, 114–15; effects of, 3, 14–15, 23–24, 32–34, 37, 43–49, 52–53, 57, 63, 67, 90, 99, 104, 108, 109–23, 128, 156–59, 173–74, 178–86 *passim*, 187; effects of 1891–93 drought on, 37–38, 109, 117; 1885 boom in, 37, 116–17; modern, 40, 114, 118, 186; repeat photography of, 66, 69, 73, 75, 79, 81, 83, 85, 87, 89. *See also* Cattle; Fire suppression; Horses; Rodents/lagomorphs
Lochiel, 145
Logging, 36–37, 166–72; amounts of timber cut, 46, 147–48, 166–72, 182–83, 184, 196; control of, 182–83; effect on vegetation, 15, 166–67, 172, 184. *See also* Woodcutting
London rocket, 155, 156
Longfellow Mine, 36
Luis María Baca land grant, 114
Lupines, 24
Lupinus spp., 24
Lycium spp., 178

McCloskey Canyon, 168
Madera Canyon, 167–68, 171
Madrean evergreen woodland. *See* Evergreen woodland
Madrone, 26
Mallows, 24
Mammoth, 36, 145, 162, 169
Manzanita, 26, 145
María Santísima del Carmen land grant, 114
Maricopa, 134
Marijilda Canyon, 171
Martynia spp., 24
Matched photography. *See* Ground photography; Aerial photography
Mendora scabra, 178
Mesquite, 26, 61, 62, 147, 165–66; changes in, 4, 24, 36, 39, 47–50, 53, 63, 66, 67, 90, 98, 99, 104, 105, 111, 119–20, 135, 138, 157, 158, 161, 174, 179, 187; control and management of, 98, 119–22, 158; effect of fire on, 139–42, 181; repeat photography of, 69, 71, 73, 81, 87; used for fodder, 112; used for fuelwood, 46, 99, 143–53 *passim*, 163, 196. *See also* Bosques; Brush invasion; Curly mesquite; Honey mesquite
Mexican blue oak, 25
Mexican pinyon, 25
Mexican white pine, 166

Mexicans, early occupation and settlement by, 32–34
Middlemarch, 145
Millet, 174
Millville, 147
Mining: copper, 36, 38, 150; effects of, 34–40 *passim*, 46, 49, 83, 87, 144–49, 153–54, 183; gold, 75; major centers of, 145; silver, 36, 38, 147–49
Mixed-conifer forest, 4, 20, 26–27, 37, 91, 121, 125, 147; fire in, 129, 135, 138, 183, 184; logging of, 144, 166, 172, 184
Mogollon Rim, 124, 125, 137
Morenci, 150, 156. *See also* Clifton-Morenci
Mormon settlements, 37
Morse (Mill) Canyon, 170, 171, 184
Mortonia scabrella, 22
Mount Graham, 18, 169, 171
Mount Lemmon, 171
Mountain mahogany, 26, 145
Mowry Mine, 34
Muhlenbergia porteri, 24
Mule Mountains, 71
Murray Springs, 63

Nacozari, Sonora, 151
National Forest Reserves, 38, 87, 182–83. *See also* Coronado National Forest; Whetstone Mountains
National High Altitude Photography Program (NHAP). *See* Aerial photography, NHAP
New York Hill, 166
New York Thicket, 166
Nogales, 40, 41, 145, 164
Nolina spp., 24
Nuttall Canyon, 171

Oak woodland, 25, 154. *See also* Evergreen woodland
Ocotillo, 22
Old Dominion Mine, 36
Olneya tesota, 22, 178
One-seed juniper, 24, 25, 26, 179

Ophir Mine, 75
Opuntia bigelovii, 22, 178
Opuntia spp., 24
Opuntia versicolor, 178
Oracle, 109, 145
Oro Blanco Mountains, 134
Otter, 4
Overgrazing. *See* Livestock grazing, effects of
Overstocking. *See* Cattle, numbers of

Pajarito Mountains, 49, 134
Palisades, 171
Paloverde, 22, 49, 61, 62, 85, 143, 178
Paloverde–sahuaro community. *See* Sonoran desertscrub
Panicum antidotale, 160
Pantano, 162
Papago (Tohono O'odham) Indians, 133, 173
Paradise, 38
Parral Mining District (Mexico), 153–54
Patagonia, 34, 62, 63, 73
Patagonia Mountains, 134
Pearce, 145, 150
Peloncillo Mountains, 148
Photography. *See* Aerial photography; Ground photography
Physiography, of southeastern Arizona, 18–19
Picea engelmannii, 27
Pima County, 37, 147, 152, 156, 164
Pima Indians, 29, 31, 32, 33, 162
Pimería Alta, 31
Pinal County, 110–13
Pinaleño (Graham) Mountains, 18, 26–27, 81, 184; forest reserve, 38, 182–83; logging in, 37, 144, 166, 169–72 *passim*, 184, 196
Pine, 25–27
Pinery Canyon, 171, 184
Pinus cembroides, 25
Pinus engelmannii, 25, 166
Pinus leiophylla, 25
Pinus strobiformis, 27
Pinyon, 119, 125, 145, 182

Pinyon–juniper woodlands, 125, 135–37, 153. *See also* Evergreen woodland
Plains bristlegrass, 25
Plains grasslands, 20, 23, 24–25, 49, 138–40, 172, 179–80. *See also* Grasslands
Plains lovegrass, 25
Platanus wrightii, 28
Ponderosa pine forest: changes in, 4, 44, 90, 91, 182–84, 186–87; description of, 20, 26–27, 59; fires in, 125, 129, 135, 137–38, 184; woodcutting in, 36–37, 144, 147–48, 153–54, 166–67, 172, 184
Population of southeastern Arizona, 34–35, 40–41
Populus fremontii. *See* Cottonwood
Populus tremuloides, 27
Portal, 63
Prairie. *See* Grasslands
Prairie dog, 4
Prescribed burning, 135–38, 140–41. *See also* Fire; Fire suppression
Precipitation, 100–4. *See also* Climate
Prosopis glandulosa. *See* Honey mesquite; Mesquite
Prosopis spp. *See* Mesquite
Prosopis velutina, 28
Pseudotsuga menziesii. *See* Douglas fir
Ptychocheilus lucius, 177
Pusch Ranch, 85
Pusch Ridge, 85

Quaking aspen, 27
Quercus arizonica, 25
Quercus emoryi, 25, 154
Quercus gambelii, 27
Quercus oblongifolia, 25
Quiburi (Santa Cruz de Terrenate), 31, 69, 176–77

Rail X Ranch, 77
Railroads, 15, 35–37, 44, 46, 116, 118, 155
Ramsey Canyon, 168, 171

Ranching. *See* Livestock grazing
Range improvement programs. *See* Livestock grazing, control of
Red brome, 141, 155–56, 160
Red Mountain, 63
Redington, 62, 162
Remote subdivisions, effects of, 40, 53, 98, 173, 178, 179, 180
Repeat aerial photography. *See* Aerial photography
Repeat ground photography. *See* Ground photography
Research Ranch, 49, 159
Reventón, 34
Rhus microphylla, 22
Rhus spp., 22, 26, 145
Rillito, 162
Rillito Creek, 30, 161, 162, 163, 174
Rincon Mountains, 26, 131–37 *passim*
Rio Grande, 50, 52, 104, 113, 195
Riparian wetlands: changes in, 4, 15, 36, 39, 43, 45–48, 50, 55, 58, 90, 121–23, 125, 143, 147, 149, 153, 155, 157–58, 161, 165–66, 176–78, 186, 195; *ciénagas* in, 4, 30, 47, 176–77, 196; decline of native phreatophytes in, 99, 122, 161, 176, 177–78; description of, 20, 27–29, 32–33, 111, 117, 176–77; repeat photography of, 71, 89; succession in, 67, 177. *See also* Arroyo cutting; Erosion; Groundwater
Roads, effects on vegetation of, 15, 34, 40, 44, 46, 47, 152, 155, 158, 161, 173, 178, 183, 184
Rock Canyon, 171, 184
Rocky Mountain (Petran) forest. *See* Mixed-conifer forest; Ponderosa pine forest; Subalpine conifer forest
Rodents/lagomorphs, effect on vegetation, 42, 57, 119, 122, 139
Rosemont, 145
Rosemont Mine, 83
Rosewood, 26
Ross Mill, 170, 171
Rucker Canyon, 171, 184

Russelville, 36, 145
Russian thistle, 156, 160, 165

Sacaton, 4, 23, 89, 111, 113, 122, 141–42, 174
Safford, 41, 162
Sagebrush, 111
Saginaw, 162
Sahuaro, 14, 22, 49, 62, 178
Saint David, 36, 37, 69, 162, 164
Salero Hacienda, 79
Salero Mine, 73
Salix spp. *See* Willow
Salsola iberica. *See* Russian thistle
Salt River Valley, 29
Saltbush, 21
Saltcedar. *See* Tamarisk
San Bernardino land grant, 114
San Bernardino Valley, 18, 32, 34, 162
San Carlos Reservoir, 39, 164
San Cayetano, Hill of, 73
San Cayetano Mountains, 73
San Francisco River, 148
San Ignacio de la Canoa land grant, 114
San Ignacio del Babocomari land grant, 114, 116
San Jose, 162
San Jose de Sonoita land grant, 73, 114
San Juan de las Boquillas y Nogales land grant, 69, 71, 114
San Manuel, 41
San Pedro, 35, 36, 162
San Pedro Riparian National Conservation Area, 69, 71
San Pedro (River) Valley, 18, 20; early Anglo settlement of, 32, 34, 37, 147; fires in, 132, 134; livestock grazing in, 112, 115–16; modern farming in, 39–40, 161–64 *passim*; prehistoric agricultural settlement in, 29–30, 38; repeat photography of, 69, 71; vegetation change in, 48–49, 59, 67, 122, 151, 158, 176–78
San Rafael de la Zanja land grant, 89, 114

San Rafael del Valle land grant, 114
San Rafael Valley, 18, 25, 89, 116, 135, 148
San Simon, 162
San Simon (Creek) Valley, 18, 19, 20, 117–18, 122–23, 128, 161–65 *passim*, 180
San Xavier (del Bac), 29, 31–34 *passim*, 36, 162, 166
Sandpaper bush, 22
Santa Catalina Mountains, 26–27; fires in, 131–35 *passim*; forest reserve, 38, 182–83; logging in, 36–37, 144, 166–67, 169, 170, 172, 184; vegetation change in, 156
Santa Cruz County, 40, 117, 130, 147
Santa Cruz de Terrenate. *See* Quiburi
Santa Cruz (River) Valley, 18, 20, 21, 89, 130, 177; description of, 40, 110–13; livestock grazing in, 112, 115–18; modern farming in, 37, 38, 39, 158, 161–65 *passim*; prehistoric agricultural settlement in, 29, 30; settlement before 1870 in, 31–35 *passim*
Santa Rita Experimental Range, 87, 120, 140
Santa Rita Mine, 73, 144
Santa Rita Mountains, 26, 75, 79, 83, 87; fires in, 131–34 *passim*; forest reserve, 38, 87, 182–83; livestock grazing in, 116, 117, 183; logging in, 36, 166–68, 170, 184; mining in, 34; settlement before 1870, 34; vegetation change in, 174, 179
Santa Teresa Mountains, 26, 144, 171
Sawmill Canyon, 167–68, 170, 171
Schismus arabicus, 155
Schismus barbatus, 141, 155
Schismus spp., 156
Scottsdale, 165
Semidesert grasslands, 20, 22, 23–24. *See also* Grasslands
Senecio, 119
Setaria macrostachya, 25
Shingle Mill, 171
Shortgrass prairie. *See* Grasslands
Sierra Bonita Ranch, 81, 109

Sierra San Jose, 179
Sierra Vista, 41, 151, 196
Sierrita Mountains, 116, 133
Silk tassel, 26
Silver Bell, 36, 145, 174-75
Silver King Mine, 36, 145, 151
Sisymbrium irio, 155, 156
Six-weeks grama. *See* Grama
Snakeweed, increase of, 4, 24, 53, 63, 67, 81, 119-20, 139, 179. *See also* Brush invasion
Sobaípuri Indians, 29, 31, 69
Soil Conservation Service: introduction of exotics by, 156, 158, 160; survey photography by, 90, 91, 96, 99
Solomonville, 169
Solomonville Drainage Ditch, 165
Sonoita (Creek) Valley, 30, 34, 37, 73, 77, 115-16, 117, 162
Sonoita, 77
Sonoita-Elgin area, 25, 162
Sonora, Mexico, 47, 48, 56, 151, 152, 154, 180
Sonoran desertscrub, 20-22, 59, 61, 91, 97-100, 141-42, 178-79. *See also* Desertscrub
Sopori, 144
Sorghum halapense, 160
Southern Pacific Railroad, 35-37, 116, 170
Southwestern white pine, 27
Spaniards, early occupation and settlement by, 30-34
Sphaeralcea spp., 24
Spiderlings, 24
Sporobolus spp. *See* Sacaton
Sporobolus wrightii. *See* Sacaton
Staghorn cholla, 178
Steam Pump Ranch, 85
Stock Raising Act of 1916, 38, 118
Stream entrenchment. *See* Arroyo cutting
Subalpine conifer forest, 20, 26-27
Subdivisions. *See* Remote subdivisions
Sulphur Springs Valley, 18, 20, 110-13, 115, 157, 162, 163, 164, 173

Sumac, 26, 145
Sunnyside Canyon, 171
Sunset, 147
Superior, 36, 38, 144
Surveyors' field notes, 15, 42, 43, 59-63, 103, 152, 173, 176, 180, 186, 196
Sycamore, 111, 176

Tamarisk: changes in, 28, 39, 53, 67, 99, 157-58, 161, 165-66, 176; and flood hazard, 157-58, 165. *See also* Riparian wetlands
Tamarix chinensis. See Tamarisk
Tanque Verde Mountains, 118
Tarbush, 22
Taylor Grazing Act, 38
Teddy bear cholla, 22, 178
Thatcher, 41, 116, 162
Threeawn, 24, 25, 174
Tidestromia spp., 24
Tobosa, 23, 24, 141, 142
Tohono O'odham Indians, 52, 133, 173
Tombstone, 35, 128, 129-31, 174, 196; fuelwood consumption in, 146-54; mining in, 38, 145, 147-48, 150; timber demand in, 168, 170, 184
Tombstone Hills, 71
Tombstone woodshed, 146-54
Tortolita Mountains, 133
Total Wreck, 145
Tres Alamos, 162
Trichachne californica, 24
Tripp Canyon, 171
Tubac, 31, 32, 33, 36, 162, 168
Tucson, 61, 85, 109, 113, 175, 196; early Anglo settlement of, 35-36; fires around, 129-37, 138, 169; fuelwood demand in, 143, 145, 151, 168; livestock grazing in, 31, 32, 33, 114-16; modern farming in, 155, 162-65; population of, 34, 41; settlement before 1870, 31-32
Tucson Plant Materials Center, 156
Tumacácori, 31, 32, 33
Tumamoc Hill, 52, 155

Tumbleweed, 156, 160, 165
Turnerville, 168

Urbanization. *See* Remote subdivisions
U.S. Public Land Survey. *See* Surveyors' field notes

Vauquelinia californica, 26
Vegetation, description of, 20–28
Vegetation change, 3, 4, 14–15, 44, 48–50, 52–56, 62–63, 66–67, 90, 98–100, 103–4, 178, 181, 185–87; causes of, 14–16, 42–58 *passim*, 67, 185; means to assess, 16, 42, 59–105; previous studies of, 42–58; succession, 67, 105, 177, 185
Velvet ash, 28. *See also* Riparian wetlands
Vine mesquite, 174
Viscid acacia, 22. *See also* Acacia

Walnut, 28. *See also* Riparian wetlands
Washington Camp, 145, 168
Webb Canyon, 171
Whetstone Mountains, 38, 117, 131, 133, 134, 147, 152, 180
White fir, 27
White ratany, 178

White-mats, 24
Whitewater Draw, 18
Wild hay harvesting, 87, 122, 128; from 1850 to 1920, 172–74, 180
Wildfire. *See* Fire
Willcox, 41, 144, 180
Willcox Playa, 18
Willow, 4, 28, 56, 66–67, 71, 89, 144, 153, 165, 176–77. *See also* Riparian wetlands
Winchester Mountains, 26
Winkelman, 145
Wolf, 4
Woodcutting: amounts cut, 145–51; by Indians, 29–30; camps, 145; effects on vegetation, 15, 43, 44, 48, 49, 66, 67, 98–99, 143, 151–54, 177–78, 180, 182, 183, 186, 195–96; estimated consumption, 145–51; for domestic uses, 38, 145–46, 149, 151–52, 154; for mining uses, 34, 36, 38, 73, 83, 143–46, 148–51, 196; management of, 38, 99, 151–53, 182. *See also* Charcoal; Logging; Wood stumps
Wood stumps, as fuel, 143, 153
Woody plants, increase of. *See* Brush invasion

Yucca, 22, 24

ABOUT THE AUTHOR

Conrad J. Bahre is a professor of geography at the University of California, Davis. Trained in cultural plant geography, he has written numerous articles and a book concerning human impacts on vegetation and vegetation change in southern Arizona, northwestern Mexico, and north-central Chile. Bahre received his B.A. and M.A. degrees from the University of Arizona and his Ph.D. from the University of California, Riverside.